T0207463

Communications in Computer and Information Science   1751

More information about this series at https://link.springer.com/bookseries/7899

Philippe Fournier-Viger · Ahmed Hassan ·
Ladjel Bellatreche · Ahmed Awad ·
Abderrahim Ait Wakrime · Yassine Ouhammou ·
Idir Ait Sadoune (Eds.)

Advances in Model and Data Engineering in the Digitalization Era

MEDI 2022 Short Papers and DETECT 2022
Workshop Papers, Cairo, Egypt, November 21–24, 2022
Proceedings

 Springer

Editors
Philippe Fournier-Viger 🆔
Shenzhen University
Shenzhen, China

Ladjel Bellatreche 🆔
ISAE-ENSMA
Poitiers, France

Abderrahim Ait Wakrime 🆔
Mohammed V University
Rabat, Morocco

Idir Ait Sadoune 🆔
University of Paris-Saclay
Gif-sur-Yvette, France

Ahmed Hassan 🆔
Nile University
Giza, Egypt

Ahmed Awad 🆔
University of Tartu
Tartu, Estonia

Yassine Ouhammou 🆔
LIAS/ISAE-ENSMA
Futuroscope, France

ISSN 1865-0929   ISSN 1865-0937 (electronic)
Communications in Computer and Information Science
ISBN 978-3-031-23118-6   ISBN 978-3-031-23119-3 (eBook)
https://doi.org/10.1007/978-3-031-23119-3

This Springer imprint is published by the registered company Springer Nature Switzerland AG
The registered company address is: Gewerbestrasse 11, 6330 Cham, Switzerland

Preface

MEDI (the International Conference on Model and Data Engineering) is an annual research conference. It focuses on advances in data management and modeling, including topics such as data models, data processing, database theory, database systems technology, and advanced database applications. MEDI was founded by researchers from Euro-Mediterranean countries and has served as a springboard for multiple international projects and collaborations. The previous editions of MEDI have been held organized in Cyprus, Estonia, France, Greece, Italy, Morocco, Portugal, and Spain.

For its 11th edition, MEDI 2022 was held in Cairo, Egypt, during November 21–24, 2022, in hybrid mode (in person and online) due to the special circumstances related to the COVID-19 pandemic. A total of 65 submissions were received, from which 18 papers were accepted for full presentation and 11 for a short presentation. This volume contains the 11 short presentation papers, as well as four accepted papers from the DETECT 2022 workshop, held at MEDI 2022. Altogether, the 15 papers included in this volume cover a wide range of topics related to the theme of the conference.

The DETECT 2022 workshop on Modeling, Verification and Testing of Dependable Critical Systems was selected after an open call for workshop proposals and the subsequent evaluation of these proposals by the MEDI workshop chair, Ahmed Awad. The DETECT workshop had its own chair, Program Committee chairs, and Program Committee.

We would like to thank everyone that contributed to the success of MEDI and DETECT 2022, including authors, organizers, and Program Committee members.

For readers, we hope that you will find useful and interesting ideas.

October 2022

Philippe Fournier-Viger
Ahmed Hassan
Ladjel Bellatreche

Organization

MEDI General Chairs

Ahmed Hassan   Nile University, Egypt
Ladjel Bellatreche   ISAE-ENSMA, France

MEDI Program Committee Chair

Philippe Fournier-Viger   Shenzhen University, China

MEDI Workshop Chair

Ahmed Awad   Tartu University, Estonia

DETECT Chair

Yassine Ouhammou   ISAE-ENSMA, France

DETECT Program Committee Chairs

Abderrahim Ait Wakrime   Mohammed V. University in Rabat, Morocco
Idir Ait Sadoune   Paris-Saclay University, France

MEDI/DETECT Proceedings Chair

Walid Al-Atabany   Nile University, Egypt

MEDI Financial Chair

Hala Zayed   Nile University, Egypt

MEDI Program Committee

Antonio Corral   University of Almeria, Spain
Mamoun Filali-Amine   IRIT, France
Flavio Ferrarotti   Software Competence Centre Hagenberg, Austria
Sofian Maabout   University of Bordeaux, France
Yannis Manolopoulos   Open University of Cyprus, Cyprus

Milos Savic	University of Novi Sad, Serbia
Alberto Cano	Virginia Commonwealth University, USA
Essam Houssein	Minia University, Egypt
Moulay Akhloufi	Université de Moncton, Canada
Neeraj Singh	University of Toulouse, France
Dominique Mery	Loria, Université de Lorraine, France
Duy-Tai Dinh	Japan Advanced Institute of Science and Technology, Japan
Giuseppe Polese	University of Salerno, Italy
M. Saqib Nawaz	Peking University, China
Jérôme Rocheteau	Icam Nantes, France
Mourad Nouioua	Hunan University, China
Ivan Luković	University of Belgrade, Serbia
Jaroslav Frnda	University of Zilina, Slovakia
Radwa El Shawi	Tartu University, Estonia
Enrico Gallinucci	University of Bologna, Italy
Anirban Mondal	The University of Tokyo, Japan
Pinar Karagoz	Middle East Technical University, Turkey
El Hassan Abdelwahed	Cadi Ayyad University, Morocco
Irena Holubova	Charles University in Prague, Czech Republic
Georgios Evangelidis	University of Macedonia, Thessaloniki, Greece
Panos Vassiliadis	University of Ioannina, Greece
Mohamed Mosbah	LaBRI, University of Bordeaux, France
Patricia Derler	Palo Alto Research Center, USA
Idir Ait Sadoune	LRI, CentraleSupélec, France
Goce Trajcevski	Iowa State University, USA
Jerry Chun-Wei Lin	Western Norway University of Applied Sciences, Norway
Yassine Ouhammou	ISAE-ENSMA, France
Srikumar Krishnamoorthy	Indian Institute of Management, Ahmedabad, India
Mirjana Ivanovic	University of Novi Sad, Serbia
Yves Ledru	Université Grenoble Alpes, France
Raju Halder	Indian Institute of Technology Patna, India
Orlando Belo	University of Minho, Portugal
Stefania Dumbrava	ENSIIE, France
Chokri Mraidha	CEA-List, France
Amirat Hanane	University of Laghoaut, Algeria
Javier Tuya	Universidad de Oviedo, Spain
Luis Iribarne	University of Almería, Spain
Elvinia Riccobene	University of Milan, Italy
Regine Laleau	Paris-East Creteil University, France

| Jaroslav Pokorný | Charles University in Prague, Czech Republic |
| Oscar Romero | Universitat Politècnica de Catalunya, Spain |

MEDI Organization Committee

Mohamed El Helw	Nile University, Egypt
Islam Tharwat	Nile University, Egypt
Sahar Selim	Nile University, Egypt
Passant El Kafrawy	Nile University, Egypt
Sahar Fawzy	Nile University, Egypt
Nashwa Abdelbaki	Nile University, Egypt
Wala Medhat	Nile University, Egypt
Heba Aslan	Nile University, Egypt
Mohamed El Hadidi	Nile University, Egypt
Mostafa El Attar	Nile University, Egypt

Abstracts of Invited Talks

A Service-based Approach to Drone Service Delivery in Skyway Networks

Athman Bouguettaya

University of Sydney, Australia
athman.bouguettaya@sydney.edu.au

Abstract. We propose a novel *service framework* to effectively provision drone-based delivery services in a skyway network. This service framework provides a high-level service-oriented architecture and an abstraction to model the drone service from both *functional* and *non-functional* perspectives. We focus on *spatio-temporal* aspects as key parameters to query the drone services under a range of requirements, including drone capabilities, flight duration, and payloads. We propose to *reformulate* the problem of drone package delivery as finding an optimal composition of drone delivery services from a designated take-off station (e.g., a warehouse rooftop) to a landing station (e.g., a recipient's landing pad). We select and compose those drone services that provide the best quality of delivery service in terms of payload, time, and cost under a range of *intrinsic* and *extrinsic* environmental (i.e., context-aware) factors, such as battery life, range, wind conditions, drone formation, etc. This talk will overview the key challenges and propose solutions in the context of single drones and swarms of drones for service delivery.

Bio: Athman Bouguettaya is Professor and previous Head of School of Computer Science, at the University of Sydney, Australia. He was also previously Professor and Head of School of Computer Science and IT at RMIT University, Melbourne, Australia. He received his PhD in Computer Science from the University of Colorado at Boulder (USA) in 1992. He was previously Science Leader in Service Computing at the CSIRO ICT Centre (now DATA61), Canberra. Australia. Before that, he was a tenured faculty member and Program director in the Computer Science department at Virginia Polytechnic Institute and State University (commonly known as Virginia Tech) (USA). He is a founding member and past President of the Service Science Society, a non-profit organization that aims at forming a community of service scientists for the advancement of service science. He is or has been on the editorial boards of several journals including, the IEEE Transactions on Services Computing, IEEE Transactions on Knowledge and Data Engineering, ACM Transactions on Internet Technology, the International Journal on Next Generation Computing, VLDB Journal, Distributed and Parallel Databases Journal, and the International Journal of Cooperative Information Systems. He is also the Editor-in-Chief of the Springer-Verlag book series on Services Science. He served as a guest editor of a number of special issues including the special issue of the ACM Transactions on Internet Technology on Semantic Web services, a special issue the IEEE

Transactions on Services Computing on Service Query Models, and a special issue of IEEE Internet Computing on Database Technology on the Web. He was the General Chair of the IEEE ICWS for 2021 and 2022. He was also General Chair of ICSOC for 2020. He served as a Program Chair of the 2017 WISE Conference, the 2012 International Conference on Web and Information System Engineering, the 2009 and 2010 Australasian Database Conference, 2008 International Conference on Service Oriented Computing (ICSOC) and the IEEE RIDE Workshop on Web Services for E-Commerce and E-Government (RIDE-WS-ECEG'04). He also served on the IEEE Fellow Nomination Committee. He has published more than 300 books, book chapters, and articles in journals and conferences in the area of databases and service computing (e.g., the IEEE Transactions on Knowledge and Data Engineering, the ACM Transactions on the Web, WWW Journal, VLDB Journal, SIGMOD, ICDE, VLDB, and EDBT). He was the recipient of several federally competitive grants in Australia (e.g., ARC), the US (e.g., NSF, NIH), Qatar (NPRP). EU (FP7), and China (NSFC). He also won major industry grants from companies like HP and Sun Microsystems (now Oracle). He is a Fellow of the IEEE, Member of the Academia Europaea (Honoris Causa) (MAE) (HON), WISE Fellow, AAIA Fellow, and Distinguished Scientist of the ACM.

Broad and Deep Learning of Big Heterogeneous Health Data for Medical AI: Opportunities and Challenges

Vincent S. Tseng

National Yang Ming Chiao Tung University, Taiwan
vtseng@cs.nctu.edu.tw

Abstract. In healthcare domains, large-scale heterogeneous types of data like medical images, vital signs, electronic health records (EHR), genome, etc., have been collected constantly, forming the valuable big health data. Broad and deep learning of these big heterogeneous biomedical data can enable innovative applications for Medical AI with rich research lines/challenges arisen. In this talk, I will introduce recent developments and ongoing projects on the topic of Medical AI, especially in intelligent diagnostic decision support and disease risk prediction by using various advanced data mining/deep learning techniques including image analysis(for medical images), multivariate time-series analysis(for vital signs like ECG/EEG), patterns mining (for EHR), text mining (for medical notes), sensory analysis (for sensory data like air quality) as well as fusion methods for integrated modelling. Some innovative applications on Medical AI with breakthrough results based on the developed techniques, as well as the underlying challenging issues and open opportunities, will be addressed too at the end.

Bio: Vincent S. Tseng is currently a Chair Professor at Department of Computer Science in National Yang Ming Chiao Tung University (NYCU). He served as the founding director for Institute of Data Science and Engineering in NYCU during 2017-2020, chair for IEEE CIS Tainan Chapter during 2013-2015, the president of Taiwanese Association for Artificial Intelligence during 2011-2012 and the director for Institute of Medical Informatics of National Cheng Kung University during 2008 and 2011. Dr. Tseng received his Ph.D. degree with major in computer science from National Chiao Tung University, Taiwan, in 1997. After that, he joined Computer Science Division of EECS Department in University of California at Berkeley as a postdoctoral research fellow during 1998-1999. He has published more than 400 research papers, which have been cited by more than 13,000 times with H-Index 60 by Google Scholar. He has been on the editorial board of a number of top journals including *IEEE Transactions on Knowledge and Data Engineering (TKDE)*, *IEEE Journal of Biomedical and Health Informatics (JBHI)*, *IEEE Computational Intelligence Magazine (CIM)*, *ACM Transactions on Knowledge Discovery from Data (TKDD)*, etc. He has also served as chairs/program committee members for a number of premier international conferences related to data mining/machine learning, and currently he is the Steering Committee Chair for *PAKDD*. Dr. Tseng has received a number of prestigious awards, including IICM Medal of Honor (2021), Outstanding

Research Award (2019 & 2015) by Ministry of Science and Technology Taiwan, 2018 Outstanding I.T. Elite Award, 2018 FutureTech Breakthrough Award, and 2014 K. T. Li Breakthrough Award. He is also a Fellow of IEEE and Distinguished Member of ACM.

Contents

Applications

DETECT Workshop: Modeling, Verification and Testing of Dependable Critical Systems

Image Processing and Diagnosis

Does Deep Learning Require Image Registration for Early Prediction of Alzheimer's Disease? A Comparative Study Using ADNI Database

Aya Gamal[1]([⊠])[iD], Mustafa Elattar[1,2][iD], and Sahar Selim[1,2][iD]

[1] Medical Imaging and Image Processing Research Group,
Center for Informatics Science, Nile University, Giza, Egypt
Ay.gamal@nu.edu.eg
[2] School of Information Technology and Computer Science,
Nile University, Giza, Egypt

Abstract. Image registration is the process of using a reference image to map the input images to match the corresponding images based on certain features. It has the ability to assist the physicians in the diagnosis and following up on the patient's condition. One of the main challenges of the registration is that it takes a huge time to be computationally efficient, accurate, and robust as it can be framed as an optimization problem. In this paper, we introduce a comparative study to investigate the influence of the registration step exclusion from the preprocessing pipeline and study the counter effect of the augmentation on the model performance. We achieved the goal of the study through three experiments using Alzheimer's Disease Neuroimaging Initiative (ADNI) dataset. Finally, a T-statistical test is applied to validate our hypothesis with a p-value of 0.027, in which case the null hypothesis should be rejected. Our proposed approach of using augmentation without any registration outperforms the other experiments for the AD vs CN task with an AUC score of 94.62%.

Keywords: Image registration · Alzheimer's disease · MRI · Medical imaging · 3D convolutional neural networks

1 Introduction

Image registration has been a crucial task and fundamental component in various medical image analysis applications over the last three decades [14]. It is defined as the process of best aligning two or more different images of the same object by finding the optimal parameters of a geometric transformation. It is also referred to as image warping, matching, or fusion.

Image registration can be used in most medical imaging applications involving disease diagnosis and monitoring, image-guided treatment delivery, and post-operative assessment after surgeries. It is widely utilized as a preprocessing tool

© The Author(s), under exclusive license to Springer Nature Switzerland AG 2022
P. Fournier-Viger et al. (Eds.): MEDI 2022, CCIS 1751, pp. 3–11, 2022.
https://doi.org/10.1007/978-3-031-23119-3_1

for medical imaging tasks such as image classification, segmentation, and object detection [23]. The goal of image registration in the medical field includes: comparing two or more images of the same modality for the same object of two or more different subjects, monitoring anatomical or functional changes of the brain over time in longitudinal studies, a fusion of anatomical images information from other modalities like Computerized Tomography (CT) or Magnetic Resonance Imaging (MRI), and Positron Emission Tomography (PET) [5].

The image registration framework is made up of the following key components: image data geometry (2D/2D, 2D/3D, 3D/3D), transformation type (linear/non-linear), similarity metric, and optimization procedure. Image registration establishes the spatial correspondence between the pixels of the source (moving) image with the ones of the target image. It can be framed as an optimization problem. The selection of the transformation model type is important and highly depends on the nature of the data to be registered. Transformations with a limited number of parameters are represented by linear transformations. They can typically be split into the rigid and affine classes [1]. Distances between the original space and the registered space are preserved during rigid transformations. They can be parametrized in a 3D space through 6 degrees of freedom (3 translations and 3 rotations) and include translations, rotations, and reflections. Affine transformations preserve straight lines and planes. These are an extension of rigid transformations and include translation, rotation, reflection, scaling, and shear. These transformations can be defined in a 3D space through 12 parameters (6 rigid, 3 scaling, and 3 shear). On the other hand, non-linear transformations have a large number of parameters. Nonlinear registration is a crucial method for aligning two dissimilar images and is frequently used in image analysis in the medical field [29], since almost all anatomical parts, or organs, of the human body, are deformable structures. The biggest benefit of non-linear transformation over linear transformation is its ability to record local deformations, leading to more precise registration [25].

The development of imaging methods brings new challenges to the field of medical image registration. As investigated in [9], the main challenge is creating more precise and effective registration systems in clinically reasonable time periods. As registration tasks take a huge time to be computationally efficient, accurate, and most importantly robust.

In most medical imaging applications, the preprocessing step is integral to the whole system. Furthermore, there are many acknowledged preprocessing pipelines including various steps such as skull-stripping, affine registration, spatial resampling, image enhancement, intensity normalization, and cropping. A widely explored system that uses the registration step in its pipeline is the early prediction of Alzheimer's Disease (AD) using structural magnetic resonance (MR) images. Image registration has been applied in this situation to precisely quantify localized variations between populations, such as those between AD patients and normal controls.

AD is considered a degenerative neurologic disorder that results in the death of brain cells and brain shrinkage. It is considered the 6th most common cause of death in the world [2]. There are three main stages of AD: cognitive normal (CN), mild cognitive impairment (MCI), and Alzheimer's disease (AD). Where MCI is an early probable indicator of Alzheimer's.

There are several studies that designed their own system for Alzheimer's stage classification by proposing a preprocessing pipeline. Most studies used the image registration step whether linear or non-linear in their pipeline. [18] used a preprocessing pipeline that processes the sMRI images by spatially normalizing using a rigid registration approach [10]. Then skull stripping and image intensities normalization, and resizing. [27] applied for linear image registration by aligning the images to a standardized template using FSL FLIRT [19]. Most previous studies use packages such as FSL [4], Statistical Parametric Mapping (SPM) [7], and FreeSurfer [3] to preprocess the data. However [21] used the Clinica software platform developed by ARAMIS Lab [2], which supports FSL, SPM and FreeSurfer. They used Clinica to register the scans to a Dartel template. Another study [28] compared two different image preprocessing procedures: a "Minimal" and a more "Extensive" procedure. "Minimal" processing included a linear registration that was performed using the SyN algorithm from ANTs [12] to register each image to the MNI space [16, 22, 30] while the "Extensive" included non-linear registration and skull-stripping. The non-linear registration was performed using the Unified Segmentation approach [11] available in SPM12 [8]. [13] used non-linear registration whereas [15] performed no registration.

In our study, we focus on designing a simple computer-aided system that differentiates between the three classes of AD by presenting a quick pipeline of the preprocessing instead of using the complex multistep pipelines as mentioned previously. We measure the effect of applying for the registration as a preprocessing step vs the effect of excluding this step from the whole pipeline on the performance of the model. We also introduce the effect of employing the augmentation technique on the performance of the model results. Finally, we recorded the time consumed at the registration step implementation.

The frame of the paper is as follows: our proposed method is illustrated in Sect. 2 with a description of the used dataset, the preprocessing pipeline, the model architecture, and the experiment's details. In Sect. 3, the analysis and the results of the experiments are reported. Finally, the conclusion part in Sect. 4.

2   Materials and Methods

We compare two different approaches to brain MRI classification with 3D convolutional neural networks. The first one is studying the effect of using a registration step in the preprocessing vs excluding this step from the pipeline, and the second one is investigating the influence of using augmentation.

2.1   Dataset

Alzheimer's disease neuroimaging initiative (ADNI) dataset (adni.loni.usc.edu) is obtained in our study. The goal of the ADNI study is to identify the earliest indicators of AD and use biomarkers from different modalities to track the progress of the disease. According to the scope of this paper, we used the structural T1 weighted MRI images with a 1.5T field strength. The data were collected from subjects diagnosed with AD, MCI, or CN. Each subject has several scans from the follow-up visits over 36 months. A total of 639 scans were employed from the baseline visit of the subjects. The distribution of the subjects over the three classes is shown in Fig. 1.

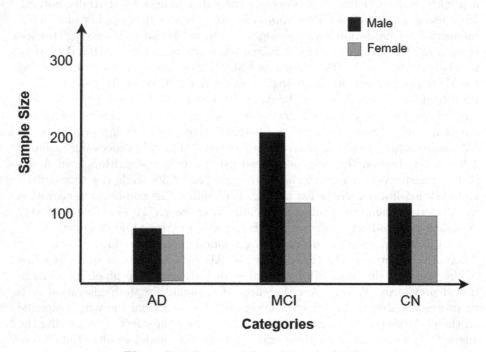

Fig. 1. Distribution of the subjects in ADNI.

2.2 Preprocessing

The selected MRIs have undergone many preprocessing steps. Firstly, a brain extraction approach is applied to the volumes to remove the skull from the whole brain using the Robust Brain Extraction (ROBEX) library [6]. For studying the effect of the registration, we then applied non-linear registration to the skull stripped images using a diffusion imaging algorithm in python (Dipy) [17] with MNI 305 atlas template. We normalized the volume intensities and applied resampling on volumes for voxel spacing. The last steps in the pipeline are to crop and resize the images.

2.3 3D Convolutional Neural Network (3D CNN)

Our 3D CNN architecture is inspired by [31]. It is composed of four convolutional layers, a global average pooling layer, and three fully connected layers. One batch normalization layer, one ReLU layer, and one max pooling layer were consecutively added to each convolutional layer. The architecture details are provided in Fig. 2.

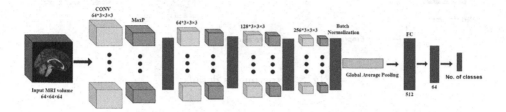

Fig. 2. 3D CNN architecture.

2.4   Experiments Details

Our experiments are framed as follows: investigating the effect of excluding the registration step from the preprocessing step versus including this step in the pipeline and measuring the augmentation influence on the performance of the whole system.

A total of 639 baseline MRI images were fed to the model after performing a 5-fold cross-validation on the dataset. Then these images were directed to different paths to achieve our experiment's structure as shown in Fig. 3. We have three directions for our experiments. The first experiment was done using minimal preprocessed images, excluding the augmentation and registration steps from the pipeline. The second experiment included the augmentation technique to apply to the preprocessed images. The last one was done on the preprocessed images including only the registrtaion step in the preprocessing pipeline. According to the scope of our study, we achieved these experiments through three different binary classification tasks: AD vs CN, AD vs MCI, and MCI vs CN. We utilized accuracy, balanced accuracy, F1-score, and area under the Receiver Operating Characteristic (ROC) curve (AUC) as evaluation metrics to evaluate the performance of our model.

We trained the three binary classification models using an Adam optimizer with a piecewise constant decay learning rate schedule, 50 epochs, and batch size 8. Finally, to validate our hypothesis, we used a T-test which is an inferential statistic used to determine whether there is a significant difference between two groups means. The T-test is applied to the AUC scores based on the folds of the results group from the registration experiment and the with-augmentation one. All these experiments are implemented using the Keras in python 3.7 on a machine with K80, P100, and T4 GPU.

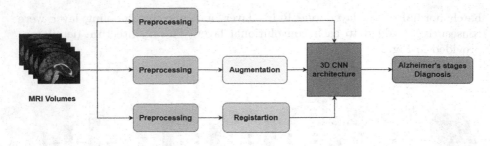

Fig. 3. Experiments details.

3   Results

The results of the experiments for the three classification tasks are reported in
Table 1 where PP Aug, PP without Aug, and PP Reg denote the experiment of
using the preprocessed images with augmentation, the experiment of using the
preprocessed images without any augmentation or registration, and the experiment of using the preprocessed images with registration respectively.

As reported in Table 1, the performance of the 3D CNN architecture when
using the preprocessed images with the augmentation only outperforms the other
experiments. It is The results of the three experiments for the AD vs CN task
are plotted in Fig. 4. Our hypothesis of the statistically significant difference that
exists between the results of the registration and with-augmentation experiments
are accepted as the resulting p-value is 0.027 from the t-test, in which case the
null hypothesis should be rejected. The comparison between the performance of
our proposed model with the previous work is shown in Table 2.

Table 1. The results of the three experiments for all tasks.

Experiment	AD vs CN			
	ACC	BA	AUC	F1-score
PP Aug	88.3	88.178	94.62	AD = 87.44,CN = 89.06
PP without Aug	73.82	70.53	77.38	AD = 60.54,CN = 80.04
PP Reg	83.5	82.01	85.42	AD = 77.18,CN = 86.92
	AD vs MCI			
	ACC	BA	AUC	F1-score
PP Aug	75.88	75.80	82.65	AD:74.28,MCI:77.02
	MCI vs CN			
	ACC	BA	AUC	F1-score
PP Aug	77.16	77.11	83.32	MCI:77.62,CN:76.54

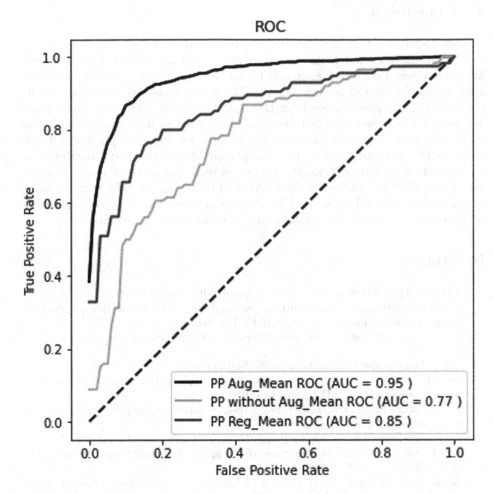

Fig. 4. Mean AUC ROC curves of the three experiments for the AD vs CN task.

Table 2. Model comparisons for the three classification tasks.

Classification task	Study	Registration	Augmentation	ACC	BA	AUC	F1-score
AD vs CN	Korolev et al. [20]	Yes	No	79	N/A	88	N/A
	Senanayake et al. [24]	No	No	76	N/A	N/A	N/A
	Vu et al. [26]	Yes	No	86.25	N/A	N/A	N/A
	PP Aug	**No**	**Yes**	**88.3**	**88.178**	**94.62**	**AD:87.44, CN:89.06**
AD vs MCI	Senanayake et al. [24]	No	No	76	N/A	N/A	N/A
	Vu et al. [26]	Yes	No	76.52	N/A	N/A	N/A
	PP Aug	No	Yes	75.88	75.80	82.65	AD:74.28, MCI:77.02
MCI vs CN	Senanayake et al. [24]	No	No	75	N/A	N/A	N/A
	Vu et al. [26]	Yes	No	85.66	N/A	N/A	N/A
	PP Aug	No	Yes	77.16	77.11	83.32	MCI:77.62, CN:76.54

4   Conclusion

In this paper, we have studied the importance of two concepts: registration and augmentation, and their effect on the performance of the whole system. We introduced a proposed approach for early diagnosis of Alzheimer's disease with augmentation and free registration. This approach achieved superior results compared to the previous work for the AD vs CN task with a minimum time. We recorded the required time to non-linear register the 3D volume with 11 min. Therefore, removing a complex step like registration from the pipeline without reducing the performance of the model is pivotal. Our proposed approach of using augmentation without any registration outperforms the previous work results for the AD vs CN task with an AUC score of 94.62%. Our hypothesis is validated using a t-test with a P-value of 0.027 which indicates that there is a significant difference between with-registration and with-augmentation results.

References

1. 153533359.pdf. https://core.ac.uk/download/pdf/153533359.pdf
2. Faststats - alzheimers disease. https://www.cdc.gov/nchs/fastats/alzheimers.htm. https://www.aramislab.fr/. Accessed 15 Aug 2022
3. Freesurfer - open source imaging. https://www.opensourceimaging.org/project/freesurfer/
4. Fsl - fslwiki. https://fsl.fmrib.ox.ac.uk/fsl/fslwiki/
5. Introduction to medical image registration. https://perso.telecom-paristech.fr/bloch/P6Image/BIOMED_Image_reg.pdf
6. Nitrc: Robust brain extraction (robex): Tool/resource info. https://www.nitrc.org/projects/robex/
7. SPM - statistical parametric mapping. https://www.fil.ion.ucl.ac.uk/spm/
8. Spm12 software - statistical parametric mapping. https://www.fil.ion.ucl.ac.uk/spm/software/spm12/
9. Alam, F., Ur Rahman, S., Hassan, M., Khalil, A.: An investigation towards issues and challenges in medical image registration. JPMI: J. Postgrad. Med. Inst. **31**(3) (2017)
10. Alansary, A., Ismail, M., Soliman, A., Khalifa, F., Nitzken, M., Elnakib, A., Mostapha, M., Black, A., Stinebruner, K., Casanova, M.F., et al.: Infant brain extraction in T1-weighted MR images using bet and refinement using LCDG and MGRF models. IEEE J. Biomed. Health Inf. **20**(3), 925–935 (2015)
11. Ashburner, J., Friston, K.J.: Unified segmentation. Neuroimage **26**(3), 839–851 (2005)
12. Avants, B.B., Epstein, C.L., Grossman, M., Gee, J.C.: Symmetric diffeomorphic image registration with cross-correlation: evaluating automated labeling of elderly and neurodegenerative brain. Med. Image Anal. **12**(1), 26–41 (2008)
13. Bäckström, K., Nazari, M., Gu, I.Y.H., Jakola, A.S.: An efficient 3D deep convolutional network for Alzheimer's disease diagnosis using MR images. In: 2018 IEEE 15th International Symposium on Biomedical Imaging (ISBI 2018), pp. 149–153. IEEE (2018)
14. Chen, X., Diaz-Pinto, A., Ravikumar, N., Frangi, A.F.: Deep learning in medical image registration. Prog. Biomed. Eng. **3**(1), 012003 (2021)

15. Cheng, D., Liu, M.: CNNs based multi-modality classification for ad diagnosis. In: 2017 10th International Congress on Image and Signal Processing, Biomedical Engineering and Informatics (CISP-BMEI), pp. 1–5. IEEE (2017)

16. Coupé, P., Manjón, J.V., Fonov, V., Pruessner, J., Robles, M., Collins, D.L.: Patch-based segmentation using expert priors: application to hippocampus and ventricle segmentation. Neuroimage **54**(2), 940–954 (2011)

17. Garyfallidis, E., et al.: Dipy, a library for the analysis of diffusion MRI data. Front. Neuroinformatics **8**, 8 (2014)

18. Hosseini-Asl, E., Gimel'farb, G., El-Baz, A.: Alzheimer's disease diagnostics by a deeply supervised adaptable 3D convolutional network. arXiv preprint arXiv:1607.00556 (2016)

19. Jenkinson, M., Bannister, P., Brady, M., Smith, S.: Improved optimization for the robust and accurate linear registration and motion correction of brain images. Neuroimage **17**(2), 825–841 (2002)

20. Korolev, S., Safiullin, A., Belyaev, M., Dodonova, Y.: Residual and plain convolutional neural networks for 3D brain MRI classification. In: 2017 IEEE 14th International Symposium on Biomedical Imaging (ISBI 2017), pp. 835–838. IEEE (2017)

21. Liu, S., Yadav, C., Fernandez-Granda, C., Razavian, N.: On the design of convolutional neural networks for automatic detection of Alzheimer's disease. In: Machine Learning for Health Workshop, pp. 184–201. PMLR (2020)

22. Louis, C.: ICBM 2009a nonlinear asymmetric MNI brain template (2015)

23. Oliveira, F.P., Tavares, J.M.R.: Medical image registration: a review. Comput. Methods Biomech. Biomed. Eng. **17**(2), 73–93 (2014)

24. Senanayake, U., Sowmya, A., Dawes, L.: Deep fusion pipeline for mild cognitive impairment diagnosis. In: 2018 IEEE 15th International Symposium on Biomedical Imaging (isbi 2018), pp. 1394–1997. IEEE (2018)

25. Song, G., Han, J., Zhao, Y., Wang, Z., Du, H.: A review on medical image registration as an optimization problem. Curr. Med. Imaging **13**(3), 274–283 (2017)

26. Vu, T.D., Ho, N.H., Yang, H.J., Kim, J., Song, H.C.: Non-white matter tissue extraction and deep convolutional neural network for Alzheimer's disease detection. Soft. Comput. **22**(20), 6825–6833 (2018)

27. Wang, H., et al.: Ensemble of 3d densely connected convolutional network for diagnosis of mild cognitive impairment and Alzheimer's disease. Neurocomputing **333**, 145–156 (2019)

28. Wen, J., et al.: Convolutional neural networks for classification of Alzheimer's disease: overview and reproducible evaluation. Med. Image Anal. **63**, 101694 (2020)

29. Ying, S., Li, D., Xiao, B., Peng, Y., Du, S., Xu, M.: Nonlinear image registration with bidirectional metric and reciprocal regularization. PLoS ONE **12**(2), e0172432 (2017)

30. Yoon, U., Fonov, V.S., Perusse, D., Evans, A.C., Group, B.D.C., et al.: The effect of template choice on morphometric analysis of pediatric brain data. Neuroimage **45**(3), 769–777 (2009)

31. Zunair, H., Rahman, A., Mohammed, N., Cohen, J.P.: Uniformizing techniques to process CT scans with 3D CNNs for tuberculosis prediction. In: Rekik, I., Adeli, E., Park, S.H., Valdés Hernández, M.C. (eds.) PRIME 2020. LNCS, vol. 12329, pp. 156–168. Springer, Cham (2020). https://doi.org/10.1007/978-3-030-59354-4_15

A CAD System for Lung Cancer Detection Using Chest X-ray: A Review

Kareem Elgohary[1,2](✉) iD, Samar Ibrahim[1,2] iD, Sahar Selim[1,2] iD, and Mustafa Elattar[1,2] iD

[1] Medical Imaging and Image Processing Research Group, Center for Informatics Science, Nile University, Sheikh Zayed, Egypt
K.Mohamed2128@nu.edu.eg
[2] School of Information Technology and Computer Science, Nile University, Sheikh Zayed, Egypt

Abstract. For many years, lung cancer has been ranked among the deadliest illnesses in the world. Therefore, it must be anticipated and detected at an early stage. We need to build a computer-aided diagnosis (CAD) system to help physicians to provide better treatment. In this study, the whole pipeline and the process of the CAD system for lung cancer detection in Chest X-ray are provided. It demonstrates the limitations and the problems facing lung cancer detection. New work is highlighted to be explored by the researchers in this area. Existing studies in the field are reviewed, including their benefits and drawbacks.

Keywords: Lung Cancer · Chest X-ray · Deep learning · CAD systems

1 Introduction

For decades, lung cancer has been one of the world's most common and deadly diseases. Every year, we have a huge number of patients diagnosed with pulmonary cancer [1]. Late diagnosis causes many patients to lose their lives by entering late cancer stages. Early detection of the lung nodules helps us to provide better treatment and save the patient's life. The initial lung screening is done using computed tomography (CT) or Chest X-Ray (CXR) which provides the radiologists with detailed information. However, the diagnosis depends greatly on the efficiency of the radiologist. This way is very time-consuming, and it does not provide an accurate way for identification of the nodules. Therefore, we are trying to demonstrate how we could develop a system that helps radiologists and doctors to identify the nodules easily, efficiently, and rapidly. This type of system is called Computer-Aided-Diagnosis (CAD). CAD helps us significantly through the analysis and the detection of the nodules within a large number of scans.

CT scans are more informative than X-ray, due to its high dimensionality (3-Dimensional) compared to X-ray (2-Dimensional). Meanwhile, CT scans have many characteristics as depicted in [2], like the amount of radiation, patient anxiety during setting inside the CT equipment, and potential financial burdens in rural populations. On the other hand, the X-ray is inexpensive, portable, smaller than the CT equipment,

P. Fournier-Viger et al. (Eds.): MEDI 2022, CCIS 1751, pp. 12–25, 2022.
https://doi.org/10.1007/978-3-031-23119-3_2

and has less radiation. All of that leads to an increase in the number of CAD systems that use CT scans compared to those that use X-ray.

Recently, Artificial intelligence (AI) empowered medical applications, especially those based on image processing. Many AI algorithms boosted the performance of image processing tasks such as classification, and segmentation. This made identification, and detection easier and more efficient. The performance of AI algorithms, especially deep learning algorithms, depends greatly on the number of images fed to the model. One of the main challenges facing medical applications is suffering from a shortage of medical images. This limitation caused researchers to tackle this issue. In 2017, the NIH clinical center released 112000 images [3]. After that, more than 755000 images were released in 2019 into 3 labeled databases, MIMIC-CXR [4], patchesadChest [5], and CheXpert [3].

All of this opens up opportunities to apply deep learning techniques, especially CNN architecture which learns from the image's features, to perform classification in the medical images field, especially chest X-Ray images for different types of diseases. There are also other tasks like segmentation, identification, and detection which provide better and efficient CAD systems by using deep learning techniques in diagnosis. Hence, researchers are working to improve these tasks using deep learning [6].

The automatic identification and detection of the lung nodules using AI is important to the field of research to provide a better CAD system. Due to the importance of this area, researchers surveyed existing studies from different perspectives. The majority of existing studies discusses the X-ray CAD system in a general way, not for lung cancer specifically, and they did not mention the challenges or the potential work of that area or the whole pipeline. Each survey paper has discussed in detail a part of the pipeline on how to build a CAD system, but not all the parts of the pipeline to give the researchers the whole view on how to build a CAD system for lung cancer detection in the CXR. In [6], the authors reviewed all the studies that used deep learning chest radiographs. They categorized the work by tasks: classification, segmentation, localization, image generation, and domain adaptation in the current state of the art of using deep learning in Chest radiographs. Meanwhile, in [7], the authors reviewed segmentation techniques and demonstrated how to use the extracted data to detect cardiopulmonary abnormalities automatically. Moreover, they listed publicly available CXR datasets.

The Existing literature went over a certain component of the pipeline for building a CAD system, but not all of the elements of the pipeline, in order to offer the research a complete view of how to develop a CAD system for lung cancer detection in the Chest X-Ray (CXR).

This study aims to provide the whole view of how to build a CAD system for lung cancer detection. The contributions of this study are listed below:

- Providing the whole pipeline and the process of the CAD system for lung cancer detection in Chest X-ray
- Demonstrating the limitations and the problems facing lung cancer detection.
- Highlighting a new work needed to be explored by the researcher in this area
- Reviewing the existing studies in the area and outlining their advantages and limitations.

● Presenting the different phases of lung cancer detection systems using different deep learning models.

The rest of the sections are divided as follows: Sect. 2 overviews the general system architecture for the detection of lung cancer from CXR. In Sect. 3, the publicly available datasets are described in detail. Section 4 provides the required preprocessing steps. Section 5 demonstrates the most used method for the image classification. The concepts of deep learning and the architectures used in different tasks to build CAD systems are overviewed in Sect. 6. Finally, the discussion on what was surveyed and the recommendations of the review are given in Sect. 7.

2   Overview of CAD System Architecture for Lung Cancer Detection from Chest X-ray

The CAD system uses chest X-ray images to identify and detect nodules. Figure 1 illustrates the schematic design of a typical lung cancer CAD system. This section provides an overview of building a CAD system and explains its steps in detail. One of the crucial steps for building a CAD system is to find a suitable dataset that can be used to train deep learning models. To improve the efficiency of the deep learning models, preprocessing of the data is required. The preprocessing steps include bone exclusion, segmentation, and nodule enhancement. Finally, the nodules would be classified using deep learning models. The phases of the CAD system and the algorithms used are presented in the next sections.

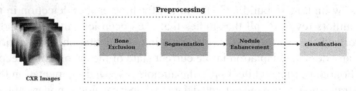

Fig. 1. General Architecture of a CAD System for lung cancer detection from chest x-ray

3   Datasets

Finding a suitable dataset is very important for training the model. Deep learning makes use of a lot of annotated data. The available datasets of the identification and prediction tasks in X-ray images are discussed below.

3.1   Japanese Society of Radiological Technology (JSRT)

The JSRT dataset contains 247 images with nodule locations besides whether the nodule is malignant or benign. Each image has a resolution of 2048 by 2048, a pixel size of 0.175 mm, and a 12-bit depth [8].

3.2 Chexpert

Chexpert is made up of 224,316 CXRs which have resolution of 8-bit grayscale. This dataset has 14 labels dividing into 12 abnormalities, no findings, and the presence of support devices [3].

3.3 National Institutes of Health Chest X-ray (NIH Chest X-ray)

There are 112,120 X-ray images in this dataset collected from 30,805 distinct patients with 14 disease labels. To create these labels, the authors used Natural Language Processing to text-mine disease classifications from associated radiological reports [3].

3.4 MIMIC-CXR

The MIMIC-CXR was originated from the Massachusetts Institute of Technology presents 371,920 chest X-rays associated with 227,943 imaging studies from 65,079 patients [5].

3.5 The Prostate, Lung, Colorectal and Ovarian (PLCO)

This dataset contains 185,421 CXRs from 56,071 patients. There are 22 disease labels with four abnormality levels and their locations. The National Institutes of Health distributes a standard set of 25,000 patients and 88,847 frontal CXRs [9].

3.6 RSNA-Pneumonia

The RSNA contains 30,000 CXRs with their annotations [10]. These images are 8-bit grayscale with a resolution of 1024×1024 and were obtained from Chest X-ray 14. The annotations were added by radiologists using bounding boxes around lung opacities. There are three classes indicating normal, lung opacity, and not normal.

3.7 National Lung Screening

This dataset is a publicly available of CXRs from 26,732 participants, and a portion of the data is available upon request. This study was aiming to compare the use of low dose computed tomography (CT) with CXRs for screening lung cancer in smokers [11].

One limitation that faces this area is that all of the above-mentioned datasets are just chest X-Ray images that don't contain annotations for lung cancer nodules, except for the JSRT dataset. To support CAD systems in this area, notated datasets must be added.

4 Preprocessing

The preprocessing phase has a great impact on the performance of the designed model. The problem of computational time and power appears due to the huge number of images and parameters. Therefore, processing the region of interests (ROIs) will improve the deep-learning algorithm's accuracy and reduce the computational time. The following subsections will discuss the main parts of the preprocessing steps for lung cancer detection.

4.1 Bone Suppression

The presence of ribs in radiographs may obscure abnormalities or nodules. This complicates the feature extraction phase of a CAD system. As a result, bone suppression is regarded as an important preprocessing step in lung segmentation and feature extraction in chest radiology. This step improves the visibility of soft tissue density and the accuracy with which abnormalities are detected. In [24], the authors found that bone suppression could increase the performance of recognition of local pneumonia significantly. The old method for removing the ribs in CXR is mainly applying Dual-Energy Subtraction (DES) imaging as proposed by [25]. The steps of DES radiography as follow: firstly, two radiographs using X-ray radiation are required: one of them at high energy and the other at low energy. After that, a subtracted image is created by combining these two radiographs using a specific weighting factor. This subtracted image highlights soft tissue or skeletal components. However, the specialized equipment is critically needed to use this technology. Hence, this DES system exists in only few hospitals. A better solution would be to detect and remove the ribs automatically, as proposed by [26]. The authors developed a technique for suppressing the contrast of clavicles and ribs in CXR by multiresolution massive training artificial neural networks. In [27], the authors proposed a method that used independent component analysis (ICA) to separate the ribs from lung images. The results showed that 90% of the ribs could be completely and partially inhibited, and 85% of the cases increased the nodule visibility. In [12], the authors analyzed the impact of bone shadow exclusion techniques in a Deep Neural Network (DNN)-based lung nodule detection algorithm. Their approach achieved higher training and validation accuracy than the image data before shadow exclusion.

4.2 Segmentation

Segmentation is one of the most important steps in the preprocessing phase. It is focused on the identification of anatomy, foreign objects, or abnormalities in the problem. There are several techniques for segmenting the lung tissue to ease the deal with nodules. These methods can be categorized into five categories as follows:

Rule-Based Method. The algorithms that are based on this method consist of a set of sequential steps that try to segment the lung. A rule-based method is generally used as an initial step before segmentation algorithms. In [13], the authors used the Euler number method to extract the lung region and refine it by morphological operations. Meanwhile, in [14], the authors applied sequential steps such as Gaussian derivative filtering, thresholding, border cleaning, noise removal, and clavicle elimination. Finally, fuzzy C-means clustering was applied. It is observed that algorithms which are based on the rule-based method are easy to implement, but they are not optimal.

Pixel Classification-Based Methods. In that group of algorithms, the pixels are labeled as lung or not by using a classifier (e.g., support vector machines, neural networks, etc....) that is trained using the images and their corresponding lung masks. In [15], the authors proposed a K-Nearest Neighbor (KNN) classifier on the pixels extracted feature. This approach takes advantage of KNN such that the posterior probability for each object class can be determined using only one neighbor search.

Model-Based Methods. The algorithms in this group use both low-level appearance and shape priors. There are mainly two types: active shape model (ASM) and Active Appearance Model (AAM). The idea behind the model-based approach is to create a model with the distribution of landmark points on training images and fit it to the test image by adjusting the distribution parameters as proposed in [16]. In [17], the authors also proposed a multiple Active Shape Model (M-ASM) by applying fifty non-uniformly distributed nodes assigned along the boundary. Each node is assigned to the corresponding pulmonary border segment. In [18], the authors used customized ASM to separate the left and right lungs. Thus, creating a function to impose distance and edge constraints.

Hybrid Methods. In this method, the best part of the schemes is combined to overcome the down flaws in each algorithm. In [19], the authors proposed a non-rigid registration-driven for lung segmentation that applies scale-invariant feature transform (SIFT-flow) for deformable registration of training masks to the patient's chest x-ray.

Deep Learning Method. Although there is popularity for using deep learning algorithms in medical imaging, there is a limited number of studies reported in the literature for lung boundary detection in CXRs. In [20], the authors proposed a semantic segmentation approach that takes the CXR image as input and outputs a map indicating the lung region probability of each pixel. In [21], the authors proposed an encoder-decoder architecture for segmenting the lung, clavicle, and heart regions. In [22], the authors proposed a new method for segmentation that could be used in the CXR called Seg-Net. Each layer in the decoder stage of SegNet corresponds to a convolutional layer at the same level, which provides a more accurate segmentation map. In [23], the authors proposed another new method for segmentation by using generative adversarial neural networks (GANs).

4.3 Nodule Candidate Detection (Nodule Enhancement)

After segmenting the lung region, the next step is to find the possible nodules in the area. There are different techniques for that task that will be reviewed. In [18], the authors applied the scale-space representation method based on Laplacian of Gaussian (LoG) filters at different scales in order to get the bright and circular region. In [17], the authors applied the watershed algorithm. Deep learning methods such as faster CNN have been widely used in CT images, but there is no report that architecture has been successfully applied for lung nodule detection in chest X-ray because the size of the original CXR image is larger than the image that has been trained on it. Furthermore, the nodules have different sizes and densities.

5   Image Classification

It refers to tasks in which a complete CXR image is analyzed to infer a category label (classification). These tasks could be done by using deep learning architectures such as AlexNet [32], VGG [33], Inception [37], ResNet [35], DenseNet models [36], Xception

[38]. In [17], the authors used Google net structure as transfer learning to detect the nodules. In [39], the authors used DenseNet-121, along with the transfer learning scheme as depicted in Sect. 6.2, to explore as a means of classifying lung cancer using chest x-ray images. The model was trained on a lung nodule dataset to identify the nodules, and then trained on the lung cancer dataset to classify the nodules as benign or malignant. This solves the issue of using a small dataset. The proposed model achieves a mean accuracy of 74.43 ± 6.01%, a specificity of 74.96 ± 9.85%, and a sensitivity of 74.68 ± 15.33%. Also, a heatmap is created by the proposed model to detect the location of the nodule. These outcomes have led to the development of the lung cancer diagnosis from chest x-rays using deep learning. Furthermore, the problem of a small dataset has been addressed. In [34], the authors proposed lung cancer and nodule prediction using ResNet-34.

6 Deep Learning for Medical Imaging

This section introduces deep learning for image analysis, focusing on the network designs that appear most frequently in the literature. Many additional studies, including a recent study of deep learning in medical image processing, provide formal definitions and more in-depth mathematical explanations of fully connected and convolutional neural networks as depicted in [28]. This study presents a cursory summary of these foundational details. The following sections describe commonly used deep learning architectures in medical image processing and provide a summary of studies that use deep learning approaches as illustrated in Table 1.

6.1 Convolutional Neural Networks (CNNs)

CNN [29] is the basis of all deep learning image tasks. Almost all deep learning image analysis tasks currently use these convolutional layers as their foundation, either as feature extractors or for classification. Neurons that connect to the previous layer are used in convolutional layers. These neurons are applied to various sections of the preceding layer, thereby acting as a sliding window over all regions and detecting the same local pattern in each position. The learned weights are exchanged while spatial information is kept.

6.2 Transfer Learning

The study of how to transfer knowledge from one domain (source domain) to another (target domain) is known as transfer learning. Pre-training is one of the most often used transfer learning procedures in CXR analysis. The pre-training strategy involves first training the network architecture on a large dataset for a distinct task and then using the trained weights as an initialization for the succeeding job for fine-tuning [30]. All layers or only the final (completely connected) layer can be re-trained, depending on the data available from the target domain. Pre-training on the ImageNet dataset has been demonstrated to be effective for chest radiography analysis [31].

6.3   Localization Network

It refers to the recognition of a specific region within an image, which is usually expressed by a bounding box or a point position. In the medical domain, localization can be utilized to detect abnormal regions, just as segmentation. The Region Convolutional Neural Network (RCNN) [40] is used to extract properties of ROI, then a support vector machine (SVM) was utilized to classify the regions. This procedure is time-consuming and comprises numerous processes. Fast-RCNN [41] shortened the processing pipeline, eliminating the need for SVM classification, and enhancing both speed and performance.

In 2017, a new extension to faster-RCNN was added to allow for more exact segmentation of the object within the bounding box. Mask R-CNN is the name of this approach [42]. Another popular object localization architecture is You Only Look Once (YOLO), which was first introduced in 2016 [43] as a single-stage object detection method. RetinaNet [44] is a single-stage detector, similar to YOLO. The majority of the localization projects in this review employ one of the architectures listed above.

6.4   Image Generation Networks

The generation of new, realistic images based on information learned from a training set is one of the tasks that deep learning is widely utilized for. In the medical domain, there are several motivations to generate images, such as improving resolution or removing projected components that impede analysis in order to produce images that are simpler to analyze. The addition of new training photos to data or the transformation of existing images to mimic appearances from a different domain (domain adaptation). Additionally, many generative techniques have been applied to enhance processes like anomaly detection and segmentation. With the development of the generative adversarial network (GAN) in 2014 [45], image generation has become more popular.

6.5   Domain Adaptation Networks

It may refer to data from a specific hardware scanner, a set of acquisition parameters, a reconstruction approach. It could also relate to demographic characteristics such as gender or even the strain of a pathogen included in the dataset, though this is less common. Domain adaption methods investigate how to accurately perform an image analysis task on data from a source domain to the target domain. Depending on the availability of labels from the target domain, these methods, which have been studied for a variety of CXR applications ranging from organ segmentation to multi-label abnormality classification, can be categorized as supervised, unsupervised, or semi-supervised. Domain adaptation techniques can be roughly separated into three divisions according to [46]: discrepancy-based, reconstruction-based, and adversarial-based. As a conclusion, there is no one-size-fits-all architecture for domain adaptation; instead, designs are mixed and matched in a variety of ways to fulfill the aim of learning to evaluate images from unknown domains.

Table 1. Summary of studies that use deep learning

Author	Model/Method
Liu et al. [47]	Segmentation-based deep fusion network (SDFN), AUC score = 0.815
Hermoza et al. [48]	Weakly supervised classification, AUC score = 0.789
Saednia et al. [49]	LSTM was used as the attention model for localization, AUC score = 0.94 Modified MobileNet architecture for classification
Yoo et al. [34]	ResNet-34 for classification, sensitivity = 0.94
Burwinkel et al. [50]	Inductive end-to-end learning strategy, in which filters from both the CNN and the graph for classification Accuracy = 0.437 ± 0.014
Wessel et al. [51]	Uses Mask R-CNN iteratively to segment and detect ribs Dice score = 0.846
Li et al. [52]	Novel CycleGAN model to decompose CXR images incorporating CT projection images for generation

7 Discussion and Conclusion

This study tried to discuss lung cancer detection in chest X-ray specifically and provide an overview of each block in the process of detection and the challenges in that field. According to the literature, there are some CAD approaches that are now being employed to detect cancer in chest radiographs. The majority of these technologies are part of the artificial intelligence area, and they focus on CAD using a chest radiograph, specifically the deep learning methods. Some of the observations made in this study are listed below:

1) It can be observed from the literature that there are different large data sets in chest X-ray in general, but not for lung cancer specifically, as illustrated in the dataset section. The only data that was found related to cancer in Chest X-Ray was the JSRT dataset. The performance of deep learning models is greatly affected by the size of the data. Therefore, the shortage of datasets caused a problem, which concluded that the lung cancer could not be detected accurately. To overcome this problem, the institute or hospital should provide more data in the future. The reviewed papers tried to overcome this problem by training a deep learning model on the large dataset of chest X-rays and trying to extract the information from CXR and building a transfer layer that is used on the JSRT data to classify the nodules.

2) Detecting lung lobes is a vital processing stage in the automated analysis of CXRs for cancer nodules. The ability to accurately localize the lung region and process only the ROI has a favorable impact on the overall performance of diagnosis/detection systems, which results in increasing their accuracy and efficiency. This study reviewed lung segmentation techniques recently used in the literature. We classified methods as shown in Fig. 2. The reviewed papers discussed that the preprocessing step should

be done to achieve better performance in detecting malignant nodules. The preprocessing could be segmentation for the lung or for the nodules, bone exclusion, or nodule enhancement.

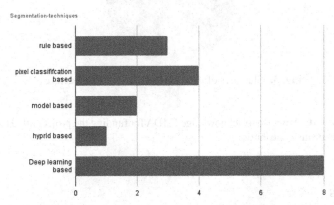

Fig. 2. Statistics of used segmentation techniques

3) Most of the reviewed work tried to perform deep learning for image tasks for better performance as shown in Fig. 3. Using deep learning for classification is the best practice to get better performance. Using the transfer learning layer as a feature extractor has a remarkable impact on the performance of the model. A large dataset such as ImageNet is used to train the model to detect nodules. The model is then fine-tuned by training it to classify the nodules in the JSRT data set. The comparison and laboratory studies given in Table 2 suggest that deep learning methods are more accurate in categorization than traditional machine learning techniques, both techniques were applied on the same data set JSRT and used the same measurement: (sensitivity) as depicted in Table 2.

Table 2. Comparison between traditional machine methods and deep learning methods

Techniques	Author	Model	Sensitivity
Machine Learning	Schilham et al. [53]	KNN	67%
	Hardie et al. [54]	Fisher linear discriminant	78.1%
Deep Learning	Wang et al. [18]	ALEXNET	69.27%
	Bush et al. [55]	RESNET	92%

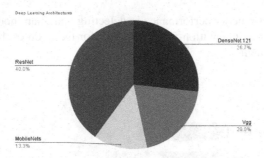

Fig. 3. Statistics of used Deep Learning Architectures

Acknowledgement. We wish to acknowledge ITIDA for funding the project titled by **CXRClear** and for their generous contributions.

References

1. Bray, F., Ferlay, J., Soerjomataram, I., Siegel, R.L., Torre, L.A., Jemal, A.: Global cancer statistics 2018: GLOBOCAN estimates of incidence and mortality worldwide for 36 cancers in 185 countries. CA: A Cancer J. Clin. **68**(6), 394–424 (2018). https://doi.org/10.3322/caac.21492
2. Fred, H.L. Drawbacks and limitations of computed tomography: views from a medical educator. Texas Heart Inst. J. **31**(4), 345–348 (2004). PMID: 15745283; PMCID: PMC548232
3. Wang, X., Peng, Y., Lu, L., Lu, Z., Bagheri, M., Summers, R.M.: ChestX-ray8: Hospital-scale Chest X-ray Database and Benchmarks on Weakly-Supervised Classification and Localization of Common Thorax Diseases. https://uts.nlm.nih.gov/metathesaurus.html
4. Johnson, A.E., et al.: MIMIC-CXR, a de-identified publicly available database of chest radiographs with free-text reports. Sci. Data **6**(1), 1–8 (2019). https://doi.org/10.6084/M9.FIGSHARE.10303823
5. Bustos, A., Pertusa, A., Salinas, J.M., de la Iglesia-Vayá, M.: PadChest: a large chest x-ray image dataset with multi-label annotated reports. Med. Image Anal. **66** (2020). https://doi.org/10.1016/J.MEDIA.2020.101797
6. Sogancioglu, E., Çallı, E., van Ginneken, B., van Leeuwen, K.G., Murphy, K.: Deep learning for chest X-ray analysis: a survey. Med. Image Anal. **72** (2021). https://doi.org/10.1016/j.media.2021.102125
7. Candemir, S., Antani, S.: A review on lung boundary detection in chest X-rays. Int. J. Comput. Assist. Radiol. Surg. **14**(4), 563–576 (2019). https://doi.org/10.1007/s11548-019-01917-1
8. Shiraishi, J., et al.: Development of a digital image database for chest radiographs with and without a lung nodule: receiver operating characteristic analysis of radiologists' detection of pulmonary nodules. AJR Am. J. Roentgenol. **174**(1), 71–74 (2000). https://doi.org/10.2214/AJR.174.1.1740071
9. Zhu, C.S., et al.: The prostate, lung, colorectal, and ovarian cancer screening trial and its associated research resource. J. Natl. Cancer Inst. **105**(22), 1684–1693 (2013). https://doi.org/10.1093/JNCI/DJT281
10. RSNA Pneumonia Detection Challenge | Kaggle. https://www.kaggle.com/c/rsna-pneumonia-detection-challenge
11. NLST - The Cancer Data Access System. https://cdas.cancer.gov/nlst/

12. Gordienko, Y., et al.: Deep learning with lung segmentation and bone shadow exclusion techniques for chest X-Ray analysis of lung cancer. In: Hu, Z., Petoukhov, S., Dychka, I., He, M. (eds.) ICCSEEA 2018. AISC, vol. 754, pp. 638–647. Springer, Cham (2019). https://doi.org/10.1007/978-3-319-91008-6_63
13. Saad, M.N., Muda, Z., Ashaari, N.S., Hamid, H.A.: Image segmentation for lung region in chest X-ray images using edge detection and morphology. In: Proceedings - 4th IEEE International Conference on Control System, Computing and Engineering, ICCSCE 2014, pp. 46–51 (2014). https://doi.org/10.1109/ICCSCE.2014.7072687
14. Wan Ahmad, W.S.H.M., Zaki, W.M.D.W., Ahmad Fauzi, M.F.: Lung segmentation on standard and mobile chest radiographs using oriented Gaussian derivatives filter. BioMed. Eng. Online 14(1), 1–26 (2015). https://doi.org/10.1186/S12938-015-0014-8/TABLES/8
15. van Ginneken, B., Stegmann, M.B., Loog, M.: Segmentation of anatomical structures in chest radiographs using supervised methods: a comparative study on a public database. Med. Image Anal. 10(1), 19–40 (2006). https://doi.org/10.1016/J.MEDIA.2005.02.002
16. Juhász, S., Horváth, Á., Nikházy, L., Horváth, G., Horváth, Á.: Segmentation of anatomical structures on chest radiographs. IFMBE Proc. 29, 359–362 (2010). https://doi.org/10.1007/978-3-642-13039-7_90
17. Chen, S., Han, Y., Lin, J., Zhao, X., Kong, P.: Pulmonary nodule detection on chest radiographs using balanced convolutional neural network and classic candidate detection. Artif. Intell. Med. 107 (2020). https://doi.org/10.1016/J.ARTMED.2020.101881
18. Wang, C., Elazab, A., Wu, J., Hu, Q.: Lung nodule classification using deep feature fusion in chest radiography. Comput. Med. Imaging Graph. 57, 10–18 (2017). https://doi.org/10.1016/J.COMPMEDIMAG.2016.11.004
19. Liu, C., Yuen, J., Torralba, A.: SIFT flow: dense correspondence across scenes and its applications. IEEE Trans. Pattern Anal. Mach. Intell. 33(5), 978–994 (2011). https://doi.org/10.1109/TPAMI.2010.147
20. Zhao, B., Feng, J., Wu, X., Yan, S.: A survey on deep learning-based fine-grained object classification and semantic segmentation. Int. J. Autom. Comput. 14(2), 119–135 (2017). https://doi.org/10.1007/s11633-017-1053-3
21. Novikov, A.A., Lenis, D., Major, D., Hladuvka, J., Wimmer, M., Buhler, K.: Fully convolutional architectures for multi-class segmentation in chest radiographs. IEEE Trans. Med. Imaging 37(8), 1865–1876 (2017). https://doi.org/10.48550/arxiv.1701.08816
22. Badrinarayanan, V., Kendall, A., Cipolla, R.: SegNet: a deep convolutional encoder-decoder architecture for image segmentation. IEEE Trans. Pattern Anal. Mach. Intell. 39(12), 2481–2495 (2015). https://doi.org/10.1109/TPAMI.2016.2644615
23. Dai, W., Dong, N., Wang, Z., Liang, X., Zhang, H., Xing, E.P.: SCAN: structure correcting adversarial network for organ segmentation in chest X-Rays. In: Stoyanov, D., et al. (eds.) DLMIA/ML-CDS -2018. LNCS, vol. 11045, pp. 263–273. Springer, Cham (2018). https://doi.org/10.1007/978-3-030-00889-5_30
24. Li, F., Engelmann, R., Pesce, L., Armato, S.G., MacMahon, H.: Improved detection of focal pneumonia by chest radiography with bone suppression imaging. Eur. Radiol. 22(12), 2729–2735 (2012). https://doi.org/10.1007/S00330-012-2550-Y
25. Vock, P., Szucs-Farkas, Z.: Dual energy subtraction: principles and clinical applications. Eur. J. Radiol. 72(2), 231–237 (2009). https://doi.org/10.1016/J.EJRAD.2009.03.046
26. Suzuki, K., Abe, H., MacMahon, H., Doi, K.: Image-processing technique for suppressing ribs in chest radiographs by means of massive training artificial neural network (MTANN). IEEE Trans. Med. Imaging 25(4), 406–416 (2006). https://doi.org/10.1109/TMI.2006.871549
27. Nguyen, H.X., Dang, T.T.: Ribs suppression in chest X-Ray images by using ICA method. In: Toi, V.V., Lien Phuong, T.H. (eds.) 5th International Conference on Biomedical Engineering in Vietnam. IP, vol. 46, pp. 194–197. Springer, Cham (2015). https://doi.org/10.1007/978-3-319-11776-8_47

28. Litjens, G., et al.: A survey on deep learning in medical image analysis. Med. Image Anal. **42**, 60–88 (2017). https://doi.org/10.1016/j.media.2017.07.005
29. Fukushima, K., Miyake, S.: Neocognitron: a self-organizing neural network model for a mechanism of visual pattern recognition, pp. 267–285 (1982). https://doi.org/10.1007/978-3-642-46466-9_18
30. Yosinski, J., Clune, J., Bengio, Y., Lipson, H.: How transferable are features in deep neural networks? In: Advances in Neural Information Processing Systems, vol. 4, pp. 3320–3328 (2014). https://arxiv.org/abs/1411.1792v1
31. Baltruschat, M., et al.: When does bone suppression and lung field segmentation improve chest X-Ray disease classification?. In: Proceedings - International Symposium on Biomedical Imaging, vol. 2019, pp. 1362–1366 (2018). https://doi.org/10.1109/ISBI.2019.8759510
32. ImageNet Classification with Deep Convolutional Neural Networks | Enhanced Reader
33. Simonyan, K., Zisserman, A.: Very deep convolutional networks for large-scale image recognition. In: 3rd International Conference on Learning Representations, ICLR 2015 - Conference Track Proceedings (2014). https://arxiv.org/abs/1409.1556v6
34. Yoo, H., Kim, K.H., Singh, R., Digumarthy, S.R., Kalra, M.K.: Validation of a deep learning algorithm for the detection of malignant pulmonary nodules in chest radiographs. JAMA Netw. Open **3**(9) (2020). https://doi.org/10.1001/JAMANETWORKOPEN.2020.17135
35. He, K., Zhang, X., Ren, S., Sun, J.: Deep residual learning for image recognition. In: Proceedings of the IEEE Computer Society Conference on Computer Vision and Pattern Recognition, vol. 2016, pp. 770–778 (2015). https://doi.org/10.1109/CVPR.2016.90
36. Huang, G., Liu, Z., van der Maaten, L., Weinberger, K.Q.: Densely connected convolutional networks. In: Proceedings - 30th IEEE Conference on Computer Vision and Pattern Recognition, CVPR 2017, vol. 2017, pp. 2261–2269 (2016). https://doi.org/10.1109/CVPR.2017.243
37. Szegedy, C., Ioffe, S., Vanhoucke, V., Alemi, A.A.: Inception-v4, inception-ResNet and the impact of residual connections on learning. In: 31st AAAI Conference on Artificial Intelligence, AAAI 2017, pp. 4278–4284 (2016). https://arxiv.org/abs/1602.07261v2
38. Chollet, F.: Xception: deep learning with depthwise separable convolutions. In: Proceedings - 30th IEEE Conference on Computer Vision and Pattern Recognition, CVPR 2017, vol. 2017, pp. 1800–1807 (2016). https://doi.org/10.1109/CVPR.2017.195
39. Ausawalaithong, W., Thirach, A., Marukatat, S., Wilaiprasitporn, T.: Automatic lung cancer prediction from chest X-ray images using the deep learning approach. In: BMEiCON 2018 - 11th Biomedical Engineering International Conference (2019). https://doi.org/10.1109/BMEICON.2018.8609997
40. Girshick, R., Donahue, J., Darrell, T., Malik, J.: Rich feature hierarchies for accurate object detection and semantic segmentation. In: Proceedings of the IEEE Computer Society Conference on Computer Vision and Pattern Recognition, pp. 580–587 (2013). https://doi.org/10.1109/CVPR.2014.81
41. Ren, S., He, K., Girshick, R., Sun, J.: Faster R-CNN: towards real-time object detection with region proposal networks. IEEE Trans. Pattern Anal. Mach. Intell. **39**(6), 1137–1149 (2015). https://doi.org/10.1109/TPAMI.2016.2577031
42. He, K., Gkioxari, G., Dollár, P., Girshick, R.: Mask R-CNN. In: IEEE Transactions on Pattern Analysis and Machine Intelligence, vol. 42, no. 2, pp. 386–397 (2017). https://doi.org/10.1109/TPAMI.2018.2844175
43. Redmon, J., Divvala, S., Girshick, R., Farhadi, A.: You only look once: unified, real-time object detection. In: Proceedings of the IEEE Computer Society Conference on Computer Vision and Pattern Recognition, vol. 2016, pp. 779–788 (2015). https://doi.org/10.1109/CVPR.2016.91

44. Lin, T.Y., Goyal, P., Girshick, R., He, K., Dollar, P.: Focal loss for dense object detection. IEEE Trans. Pattern Anal. Mach. Intell. **42**(2), 318–327 (2017). https://doi.org/10.1109/TPAMI. 2018.2858826

45. Goodfellow, I., et al.: Generative adversarial networks. Commun. ACM **63**(11), 139–144 (2014). https://doi.org/10.1145/3422622

46. Wang, M., Deng, W.: Deep visual domain adaptation: a survey. Neurocomputing **312**, 135–153 (2018). https://doi.org/10.1016/J.NEUCOM.2018.05.083

47. Liu, H., Wang, L., Nan, Y., Jin, F., Wang, Q., Pu, J.: SDFN: segmentation-based deep fusion network for thoracic disease classification in chest X-ray images. Comput. Med. Imaging Graph.: Official J. Comput. Med. Imaging Soc. **75**, 66–73 (2019). https://doi.org/10.1016/J. COMPMEDIMAG.2019.05.005

48. Hermoza, R., Maicas, G., Nascimento, J.C., Carneiro, G.: Region proposals for saliency map refinement for weakly-supervised disease localisation and classification. In: Martel, A.L., et al. (eds.) MICCAI 2020. LNCS, vol. 12266, pp. 539–549. Springer, Cham (2020). https:// doi.org/10.1007/978-3-030-59725-2_52

49. Saednia, K., Jalalifar, A., Ebrahimi, S., Sadeghi-Naini, A.: An attention-guided deep neural network for annotating abnormalities in chest X-ray images: visualization of network decision basis. In: Proceedings of the Annual International Conference of the IEEE Engineering in Medicine and Biology Society, EMBS, vol. 2020, pp. 1258–1261 (2020). https://doi.org/10. 1109/EMBC44109.2020.9175378

50. Burwinkel, H., et al.: Adaptive image-feature learning for disease classification using inductive graph networks. In: Shen, D., et al. (eds.) MICCAI 2019. LNCS, vol. 11769, pp. 640–648. Springer, Cham (2019). https://doi.org/10.1007/978-3-030-32226-7_71

51. Wessel, J., Heinrich, M.P., von Berg, J., Franz, A., Saalbach, A.: Sequential Rib Labeling and Segmentation in Chest X-Ray using Mask R-CNN (2019). https://arxiv.org/abs/1908. 08329v1

52. Li, Z., Li, H., Han, H., Shi, G., Wang, J., Zhou, S.K.: Encoding CT anatomy knowledge for unpaired chest X-ray image decomposition. In: Shen, D., et al. (eds.) MICCAI 2019. LNCS, vol. 11769, pp. 275–283. Springer, Cham (2019). https://doi.org/10.1007/978-3-030-32226-7_31

53. Schilham, A.M., van Ginneken, B., Loog, M.: A computer-aided diagnosis system for detection of lung nodules in chest radiographs with an evaluation on a public database. Med. Image Anal. **10**(2), 247–258 (2006). https://doi.org/10.1016/J.MEDIA.2005.09.003

54. Hardie, R.C., Rogers, S.K., Wilson, T., Rogers, A.: Performance analysis of a new computer aided detection system for identifying lung nodules on chest radiographs. Med. Image Anal. **12**(3), 240–258 (2008). https://doi.org/10.1016/J.MEDIA.2007.10.004

55. Bush, I.: Lung nodule detection and classification (2016). http://cs231n.stanford.edu/reports/2016/pdfs/313_Report.pdf

Machine Learning and Optimization

Enhanced IoT Based IDS Driven by Binary Snake Optimizer for Feature Selection

Ayman A. El-Saleh[1], Thaer Thaher[2,3]([✉]), Hamouda Chantar[4],
and Majdi Mafarja[5]

[1] Department of Electronics and Communication Engineering, A'Sharqiyah
University, Ibra 400, Oman
[2] Department of Computer Systems Engineering, Arab American University,
Jenin, Palestine
thaer.thaher@aaup.edu
[3] Information Technology Engineering, Al-Quds University, Jerusalem, Palestine
thaher.thaer@students.alquds.edu
[4] Faculty of Information Technology, Sebha University, Sebha, Libya
ham.Chantar@sebhau.edu.ly
[5] Department of Computer Science, Birzeit University, Birzeit, Palestine
mmafarja@birzeit.edu

Abstract. Connected devices have extended the borders of the traditional Internet into the new Internet of Things (IoT). IoT holds a significant role in several fields such as industry, transportation, smart homes, cities, and others. However, protecting IoT environments and preventing intrusions is one of the critical problems in IoT. An intrusion detection system (IDS) aims to identify malicious patterns and threats that traditional security countermeasures cannot detect. This paper presents an effective feature selection (FS) approach driven by a binary variant of the newly proposed Snake Optimizer (SO) for enhancing intrusion detection systems. Two variants of FS are developed, and the best one that is based on a V-shaped transfer function is compared with several optimization algorithms to confirm its efficiency in boosting IDSs. In addition, five datasets that represent real IoT traffic are employed for evaluation purposes. The experimental results show that SO based on V-shaped transfer is superior to the S-shaped transfer function and outperforms other optimizers in particular based on the obtained average accuracy and convergence rates. Hence, it can conclude that the proposed approach can be efficiently employed in IoT intrusion detection systems.

Keywords: Internet of things · Snake optimizer · Feature selection · Attacks detection · Classification · Machine learning

Due to the rapid expansion of the amount of IoT devices, IoT security has gained high consideration. IoT is a network of devices/things connected via the

Supported by organization x.

internet. These devices employ embedded sensors to gather information regarding the nearby environment and interact with each other for data interchange, process, and storage [1,3]. The IoT establishes a significant influence on the economy via remodeling many organizations into digital businesses and simplifying modern business models, increasing efficiency and supporting the involvement of employees and customers [3]. Currently, IoT devices are spread in a large number of applications such as healthcare, military, industry, and transportation [2]. The computational capabilities of IoT devices are limited, putting them under the threat of continuous attacks [4]. Overall, IoT is subject to different types of attacks such as sinkhole attacks, selective forwarding attacks, hello flooding attacks, wormhole attacks, Sybil attacks, and DoS attacks [5]. Generally, relying on conventional security techniques such as authentication and encryption is not sufficient for preventing attacks targeting IoT systems. Subsequently, reliable techniques that provide high security are required. Intrusion Detection Systems (IDSs) is one of the popular techniques for maintaining the security of such systems [8]. IDSs analyze the network traffic to detect potential attacks or malicious activities. Once an attack or malicious activity is identified, a warning is sent to the decision-making system by the IDSs to perform an appropriate response [8]. Ideally, according to the detection policy, IDSs are categorized into different classes comprising rule-based, data mining and machine learning based, and statistical-based approaches [8].

Previous studies show considerable interest in applying data mining and machine learning techniques in the IoT domain. Although, processing IoT data for predicting traffic or detecting suspected activities such as frauds, anomalies, and outliers are considered a highly complex problem due to the nature of IoT data. IoT data is high-dimensional and is typically presented by high dimensions (e.g., hundreds to thousands) [2]. In addition to valuable and informative features or patterns, high dimensions data often include noisy, irrelevant, and redundant features or attributes that hurt the accuracy of the machine learning model [10]. When the dimension of data is very high, machine learning classifiers need large-size training data, leading to the well-known problem in machine learning which is the curse of dimensionality [11]. This problem can be settled via feature selection (FS). FS process examines the quality of all features in the dataset to identify and select the most informative features and remove other irrelevant and noisy ones from the feature space. FS becomes an essential pre-processing step in data mining for obtaining reliable and accurate classification models. In the FS procedure, assessing the quality of the chosen subset of features can be achieved using the wrapper or filter approaches. Wrapper-based FS approaches employ an induction (learning) algorithm to identify the best feature combinations for the classification task. On the other hand, filter approaches perform the feature selection without assistance from the induction algorithm, where statistical methods such as information gain or chi-square are used to measure the quality of selected features [12,13]. Metaheuristics have gained their popularity in FS field due to their global search capabilities and exceptional performance [10,12].

Snake Optimizer (SO) is a novel meta-heuristic optimization algorithm introduced by [17] in 2022. SO simulates the unique mating behaviors of snakes in nature. Every snake (male or female) fights to possess the best partner in case the available amount of food is sufficient and the level of temperature is low [17]. SO algorithm mathematically imitates and models these mating and foraging behaviors and provides a simple and effective optimization algorithm. This paper aims to boost the performance of anomaly identification or traffic classification into attacks or normal for IoT environment by dealing with the curse of dimensionality problem in IoT traffic where an effective wrapper feature selection approach based on the new Snake Optimizer (SO) meta-heuristic algorithm is proposed. Since the SO algorithm was initially developed to solve continuous optimization problems; hence, the authors modified the main version of OS to deal with binary problems and applied it to feature selection in IoT. Two types of transfer functions, including V-shaped and S-shaped transfer functions, have been integrated into OS. Several experiments were conducted on five real-IoT datasets obtained from [2] to examine the efficiency of the proposed OS-based FS approaches in the IoT field. Several optimization algorithms have been extensively employed for FS problem [18]. However, SO has not been utilized in the area of FS yet.

The rest of the paper is structured as follows. Section 1 provides a review of related works. Section 3 illustrates the proposed methodology. Experimental results and analysis are discussed in Sect. 4. Lastly, Sect. 5 presents conclusion and future directions.

1   Review of Related Works

With the wide spread of IoT devices, security concerns in IoT have attracted many researchers to concentrate on developing effective techniques for IoT security. Machine learning and feature selection have been effectively applied for malicious traffic detection in IoT networks. For instance, a network anomaly detection approach based on fuzzy logic and genetic algorithm was proposed by [19]. Using an approach based on Multilayer Perceptron trained by artificial bee colony algorithm, [20] proposed an intrusion Detection system (IDS) that classifies network traffic into normal or malicious. An IDS system that uses Bayesian networks and C4.5 was proposed by [21] to classify the network traffic where the firefly optimization algorithm was applied to perform feature selection. To detect potential attacks, a supervised IDS system based on a fast learning network and particle swarm optimization (PSO) algorithm was developed by [22]. An artificial neural network (ANN) classifier was employed by [23] for intrusion detection and determining Distributed Denial of Service (DDoS) attacks. Experimental results demonstrated 99.4% accuracy and confirmed that the proposed system could successfully identify various DDoS/DoS attacks. In [24], a layered approach was proposed for identifying network intrusions with the help of specific rule learning classifiers. Every layer is modeled to identify an attack type by using specific nature-inspired search approaches such as genetic search, ant search, and PSO. The proposed model was evaluated using several measures such as accuracy, efficiency, false alarm rate, and detection rate. In [25], several machine learning algorithms, including decision tree, random forest, and gradient-boosting machine (GBM), were applied to analyze and predict network attacks on IoT devices. For tracking attacks, the hiring of fog-to-things architecture was addressed by [26]. An open-access dataset was employed to compare deep and shallow neural networks. The experimental test revealed that the deep network model achieved 98.27% accuracy, whereas the shallow neural network model recorded 96.75% accuracy. As presented in [27], an IoT-based intrusion-detection system was suggested where several machine learning (ML) classifiers were applied and successfully identified network plain and scanning probing sorts of denial of service (DoS) attacks.

Recently, evolutionary algorithms have been employed to develop robust feature selection approaches for many classification tasks [12]. The literature reveals that employing evolutionary algorithms for protecting IoT and developing wrapper FS approaches to improve attack detection in IoT is still limited. However, a wrapper-based FS approach for IoT intrusion detection systems that use the Bat algorithm with Swarm Division and Binary Differential Mutation was proposed by [29] for selecting the most relevant features. As presented by [2], a new wrapper feature selection technique augmented with Whale Optimization Algorithm (WOA) was developed for IoT intrusion detection systems to deal with the problem of high dimensionality. V-shaped and S-shaped transfer functions were embedded into WOA to convert it from continuous to binary. In addition, five datasets were formed from a dataset obtained from the UCI repository and used for the evaluation. The experimental results revealed that WOA based on a V-shaped transfer function integrated with an elitist tournament binarization approach is better than WOA based on an S-shaped transfer function and also outperformed other evolutionary optimizers in terms of different measures, including average accuracy, fitness, number of features, convergence curves and running time.

2   Snake Optimizer

Snake Optimizer (SO) is a novel optimization algorithm proposed by [17]. SO simulates the mating behaviors of snakes in nature. The mathematical model of the SO algorithm is as the following [17].

- *Initialization*: SO begins with a population of randomly generated individuals. The following equation is used to generate the initial population.

$$X_i = X_{min} + r \times (X_{max} - X_{min}) \tag{1}$$

where X_i denotes the position of ith member, r is a randomly generated number between 0 and 1, and $Xmin$ and $Xmax$ represent the lower and upper boundaries of the problem respectively.

– *Diving the swarm into two equal groups males and females:* It is assumed that the number of males and females are equal. Hence, The Eqs. 2 and 3 are used to divide the population into two groups.

$$N_m \simeq \frac{N}{2} \tag{2}$$

$$N_f = N - N_m \tag{3}$$

where N denotes the number of individuals, N_m indicates the number of males while N_f refers to the number of females.

– *Evaluate each group, Defining Temperature and Food Quantity:*
  • Determine the best member in each group and find the best male (f_{best}, m) and best female (f_{best}, f) and the position of food (f_{food})
  • The temperature $Temp$ is defined as following:

$$Temp = exp(\frac{-t}{T}) \tag{4}$$

  where t indicates the current generation and T denotes to the maximum number of generations.
  • Defining food quantity (Q): The food quantity is given as follows:

$$Q = c_1 * exp(\frac{t - T}{T}) \tag{5}$$

  where c_1 is a constant and its value is equal to 0.5.

– *Exploration Phase (no Food):* If $Q <$ Threshold (Threshold $= 0.25$) then the snakes look for food by choosing any random positions, and update their current positions according to the selected positions. The exploration phase can be modelled as the follows:

$$X_{i,m}(t + 1) = X_{rand,m}(t) + c_2 \times A_m \times ((X_{max} - X_{min}) \times rand + X_{min}) \tag{6}$$

where Xi, m indicates the position of the i_{th} male, $X_{rand,m}$ denotes position of a random male, rand is a randomly generated number between 0 and 1, c_2 presents a constant and equals 0.05 and A_m refers to the ability of the male to find the food, and can be estimated as follows:

$$A_m = exp(\frac{-f_{rand,m}}{f_{i,m}}) \tag{7}$$

where $f_{rand,m}$ denotes the fitness of $X_{rand,m}$ and $f_{i,m}$ is the fitness of i_{th} member in the males group.

$$X_{i,f}(t + 1) = X_{rand,f}(t) \pm c_2 \times A_f \times ((X_{max} - X_{min}) \times rand + X_{min}) \tag{8}$$

where Xi, f indicates the position of the i_{th} female, $X_{rand,f}$ denotes position of a random female, rand is a randomly generated number between 0 and 1,

c_2 presents a constant and equals 0.05 and A_f is the ability of the female to find the food, and can be estimated as follows:

$$A_f = exp(\frac{-f_{rand,f}}{f_{i,f}}) \tag{9}$$

where $f_{rand,f}$ denotes the fitness of $X_{rand,f}$ whereas $f_{i,f}$ is the fitness of i_{th} member in the females group.

– Exploitation Phase (Food Exists): If $Q >$ threshold and temperature > 0.6 (hot) then the snakes will crawl to the food only. This can be modelled as follows:

$$X_{i,j}(t+1) = X_{food} \pm c_3 \times Temp \times rand \times (X_{food} - X_{i,j}(t)) \tag{10}$$

where $X_{i,j}$ denotes the position of a member (male or female), X_{food} means the position of the best members in the population, while c_3 is a constant and equals to 2.

If the temperature $<$ threshold (0.6) (cold), that means the snake is in the mating or fighting mode. Fight mode can be modelled as follows:

$$X_{i,m}(t+1) = X_{i,m}(t) \pm c_3 \times FM \times rand \times (X_{best,f} - X_{i,m}(t)) \tag{11}$$

where $X_{i,m}$ denotes the i_{th} male position, $X_{best,f}$ indicates the position of the best member in female group, and FM denotes the fighting ability of a male individual.

$$X_{i,f}(t+1) = X_{i,f}(t) \pm c_3 \times FF \times rand \times (X_{best,m} - X_{i,f}(t+1)) \tag{12}$$

where $X_{i,f}$ denotes the i_{th} female position, $X_{best,m}$ indicates the position of the best individual in male group, and FF denotes the fighting ability of a female individual. The following equations are used to calculate FM and FF:

$$FM = exp(\frac{-f_{best,f}}{f_i}) \tag{13}$$

$$FF = exp(\frac{-f_{best,m}}{f_i}) \tag{14}$$

where $f_{best,f}$ denotes the fitness of the best individual in the female group, $f_{best,m}$ represents the fitness of the best member in the male group, and f_i is the fitness of the i_{th} individual. Mating mode in SO can be modelled as follows:

$$X_{i,m}(t+1) = X_{i,m}(t) \pm c_3 \times M_m \times rand \times (Q \pm X_{i,f}(t) - X_{i,m}(t)) \tag{15}$$

$$X_{i,f}(t+1) = X_{i,f}(t) \pm c_3 \times M_f \times rand \times (Q \pm X_{i,m}(t) - X_{i,f}(t)) \tag{16}$$

where $X_{i,f}$ denotes the position of i_{th} individual in the females group, $X_{i,m}$ is the position of i_{th} individual in the males group. M_m and M_f denotes the mating ability of males and females respectively. They can be estimated as following:

$$M_m = exp(\frac{-f_{i,f}}{f_{i,m}})$$ (17)

$$M_f = exp(\frac{-f_{i,m}}{f_{i,f}})$$ (18)

If an egg hatched, then select worst male and female and replace them using the following equations:

$$X_{worst,m} = X_{min} + rand \pm (X_{max} - X_{min})$$ (19)

$$X_{worst,f} = X_{min} + rand \pm (X_{max} - X_{min})$$ (20)

where $X_{worst,m}$ represents the worst individual in the males group while $X_{worst,f}$ denotes the worst individual in the females group.

3 The Proposed Approach

3.1 Feature Selection Using Binary Snake Optimizer

SO algorithm was originally designed to tackle continuous optimization problems. To deal with binary problems, it must be converted into binary. Transfer functions (TFs) are the most popular applied methods for this conversion. TFs are classified according to their shape into two main branches: S-shaped and V-shaped functions [28]. The sigmoid function, which belongs to S-shaped family, was originally introduced by [30] to propose the first binary variant of PSO using Eq. (21), while [31] used tanh (V-shaped) for binarizing GSA algorithm Eq. (22).

$$T(x_j^i(t)) - \frac{1}{1 + e^{-x_j^i(t)}}$$ (21)

$$T(x_j^i(t)) = |\tanh(x_j^i(t))|$$ (22)

where x_i^j refers to the jth dimension of the ith individual (solution) at a specific iteration t, and $T(x_i^j(t))$ is the probability value calculated by TF. For preforming the binarization step, various methods have been incorporated with the aforementioned rules. The first binarization is the standard method as shown in Eq. (23) used by [30] with the S-shaped TF. The second binarization method (as in Eq. 24) is called complement and was used by [31] with the V-shaped TF.

$$X_i^k(t+1) = \begin{cases} 1 & r < T(x_j^i(t))) \\ 0 & Otherwise \end{cases}$$ (23)

$$X_i^k(t+1) = \begin{cases} \sim X_i^j(t) & r < T(x_j^i(t))) \\ X_i^j(t) & Otherwise \end{cases}$$ (24)

where r is a random number in $[0, 1]$ interval, $\sim$ denotes the complement, and $X_i^j(t+1)$ presents the binary output. In this work, these two TFs were employed to transform the original version of Snake Optimizer into binary versions, and applied them for feature selection in IoT.

3.2   Fitness Function

A practical fitness function is needed to guide the search process. Hence each generated subset of features is assigned a score to determine its quality. The main objective of the FS process is to maximize the classification accuracy and decreases the number of selected features. These two contradictory measures can be formulated using Eq. (25)

$$\downarrow Fitness = \alpha \times Er + \beta \times \frac{|S|}{|T|} \tag{25}$$

where Er means the classification error rate, $|S|$ refers to the number of chosen features and $|T|$ indicates the number of all features. α and its complement β present controlling parameters $\in [0, 1]$ that are employed to obtain the desired balance between both measures.

4   Experimental Results

4.1   Datasets and Parameter Settings

In this work, five datasets are employed to test the efficiency of the proposed approach in classifying IoT data. These datasets were obtained from [2]. Table 2 presents the description of the datasets. They were originally sampled from a dataset set obtained from the UCI repository.

The proposed FS method was implemented using MATLAB-R2018a, and the KNN classifier (k = 5) [2] was employed as an internal evaluator due to its simplicity and low cost of computation compared to its peers. All experiments were performed on a computer with Intel(R) Core(TM) i7-1165G7 CPU @ 2.80 GHz (8 CPUs) and 16 GB RAM. To be fair and consistent, all approaches in this work were tested using the same common parameters (e.g., 100 iterations and population size of 10). These values were set after performing a set of extensive experiments. Parameters of tested algorithms were decided based on recommended settings in the original papers as well as related works in FS. Table 1 provides the list of parameters and their values.

Table 1. The employed parameters settings

Fitness function	$\alpha = 0.99$, $\beta = 0.01$
Common parameters	No. of iterations = 100, population size = 10, No. of runs = 20
Classification	K = 5 (KNN classifier) with 10-fold cross validation
GSA	$G_0 = 100$, $\alpha = 20$
BBA	$Q_{min} = 0.0$, $Q_{max} = 2.0$, r pulse rate = 0.5, A loudness = 0.5
GA	Mutation ratio = 0.1 and Crossover ratio = 0.9, Roulette Wheel selection

Table 2. Description of training and testing sets including the distribution of normal and attacks cases (packets) for the training & testing parts

	Training dataset			Testing dataset			
	Normal traffic	Attack traffic	Types of attacks	Normal traffic	Attack traffic	Normal to attack ratio	Types of attacks
Dataset1	1664	1664	{COMBO, UDP}	128	1600	1:13	All (10 types)
Dataset2	1664	1664	{TCP, UDP}	128	1600	1:13	All (10 types)
Dataset3	1664	1664	{SCAN, SYN}	128	1600	1:13	All (10 types)
Dataset4	1664	1664	{UDP, ACK}	128	1600	1:13	All (10 types)
Dataset5	1664	1664	{TCP, UDPPLAIN}	128	1600	1:13	All (10 types)

4.2 Results of BSO Based S-shaped and V-shaped TFs

Inspecting Table 3, it is evident that the V-shaped transfer function based FS is better than the S-shaped transfer function based FS approach over all five examined datasets in terms of all considered measures, including classification accuracy, number of selected features, fitness and time of computation (denoted in bold). This performance improvement could be interpreted by the sharp flipping between 0 and 1 when using the V-shaped function, which emphasizes the exploration behaviors of the SO algorithm.

Table 3. Comparison between two versions of BSO with S-shaped and V-shaped TFs

Dataset	Mesure	Accuracy		Number of Features		Fitness		Time	
		SBSO	VBSO	SBSO	VBSO	SBSO	VBSO	SBSO	VBSO
Data1	AVG	0.98959	**0.98964**	58.70	**48.10**	0.0154	**0.0144**	627.93	**481.83**
	STD	0.0031	0.0031	4.6200	6.9833	0.0029	0.0030	5.4071	10.6841
Data2	AVG	0.99855	**0.99849**	66.10	**45.70**	0.0072	**0.0055**	631.87	**483.20**
	STD	0.0007	0.0005	8.5173	9.1900	0.0011	0.0009	3.8779	8.1353
Data3	AVG	0.95463	**0.96065**	57.90	**48.40**	0.0500	**0.0432**	625.07	**482.94**
	STD	0.0029	0.0049	3.8137	6.9952	0.0027	0.0049	3.2224	10.7459
Data4	AVG	0.98843	**0.98843**	53.70	**40.10**	0.0161	**0.0149**	613.72	**487.46**
	STD	0.0000	0.0000	1.7670	7.1251	0.0002	0.0006	4.2688	13.1576
Data5	AVG	0.97164	**0.97164**	63.40	**43.20**	0.0336	**0.0318**	624.12	**473.71**
	STD	0.0030	0.0030	9.1190	8.6513	0.0025	0.0034	6.1243	9.8778
Ranking	W|T|L	0|0|5	**5|0|0**	0|0|5	**5|0|0**	0|0|5	**5|0|0**	0|0|5	**5|0|0**

4.3 Comparison of VBSO with Other Well-Known Optimizers

In this section, we present a comparison between the proposed VBSO and other well-established algorithms in the literature. For this purpose, we utilized binary versions of bat algorithm (BBA), gravitational search algorithm (BGSA), salp swarm algorithm (BSSA), ant-lion optimizer (BALO), and genetic algorithm (GA). To conduct a fair comparisons, all FS approaches were implemented, and all trials were conducted under the same environment using the same parameter settings. In terms of classification accuracy, as shown in Table 4, it can be observed that the VBSO approach outperformed all its peers over all examined datasets except in the case of Data4, where the BALO FS approach has recorded the same accuracy rate as VBSO. In terms of the lowest number of selected features, as presented in Table 5, it is clear that GA is the best compared to all other considered algorithms. However, VBSO achieved a lower average of selected features than BGSA, BALO, and BSSA.

Table 6 provides a comparison of VBSO with other well-known optimizers in terms of fitness values. It can be seen that VBSO is superior compared to all other algorithms. In addition, Table 7 reveals that the GA algorithm is dominant in terms of the lowest average of computational time. However, VBSO recorded a lower average computation time than three algorithms, including BGSA, BALO, and BSSA. Furthermore, Fig. 1 shows convergence trends of all considered algorithms on the five utilized datasets. It is clear that the BSO achieved the best performance among other algorithms, where it recorded the fastest convergence rates and the lowest fitness values in all tested cases. However, it can be seen that other algorithms such as BGSA and BSSA suffer from premature convergence problems in some cases.

Table 4. Comparison of VBSO with other well-known optimizers in terms of accuracy rates

Benchmark	Measure	BGSA	BALO	BBA	BSSA	GA	VBSO
Data 1	AVG	0.9649	0.9727	0.9309	0.9566	0.9726	**0.9896**
	STD	0.0192	0.0110	0.0530	0.0149	0.0120	0.0031
Data 2	AVG	0.9919	0.9912	0.9159	0.9898	0.9901	**0.9985**
	STD	0.0053	0.0053	0.1036	0.0047	0.0046	0.0005
Data 3	AVG	0.9386	0.9537	0.8612	0.9491	0.9516	**0.9606**
	STD	0.0200	0.0000	0.1075	0.0065	0.0069	0.0049
Data 4	AVG	0.9717	**0.9884**	0.9085	0.9813	0.9792	**0.9884**
	STD	0.0296	0.0000	0.0964	0.0116	0.0120	0.0000
Data 5	AVG	0.9653	0.9678	0.8387	0.9677	0.9672	**0.9716**
	STD	0.0021	0.0007	0.1383	0.0008	0.0020	0.0030
Ranking	F-Test	4.2	2.1	6	4	3.6	**1.1**

Table 5. Comparison of VBSO with other well-known optimizers in terms of number of selected features

Benchmark	Measure	BGSA	BALO	BBA	BSSA	GA	VBSO
Data 1	AVG	53.00	71.20	47.20	64.70	**22.60**	48.10
	STD	3.8297	15.1203	8.3373	15.3337	3.3731	6.9833
Data 2	AVG	50.70	66.60	44.80	58.00	**23.70**	45.70
	STD	5.4579	13.2346	6.8118	9.2496	2.7101	9.1900
Data 3	AVG	53.90	75.00	43.40	62.40	**21.50**	48.40
	STD	3.8137	11.2842	7.3967	8.7076	3.7786	6.9952
Data 4	AVG	52.40	63.90	46.50	58.70	**22.40**	40.10
	STD	4.9710	11.0096	6.6875	10.0338	2.1705	7.1251
Data 5	AVG	51.10	63.50	45.80	65.80	**23.00**	43.20
	STD	3.3483	13.5175	9.0406	11.7644	3.2318	8.6513
Ranking	F-Test	4	5.8	2.4	5.2	1	2.6

Table 6. Comparison of VBSO with other well-known optimizers in terms of fitness values

Benchmark	Measure	BGSA	BALO	BBA	BSSA	GA	VBSO
Data 1	AVG	0.0393	0.0332	0.0324	0.0486	0.0291	**0.0144**
	STD	0.0190	0.0116	0.0145	0.0141	0.0117	0.0030
Data 2	AVG	0.0124	0.0145	0.0139	0.0152	0.0119	**0.0055**
	STD	0.0051	0.0042	0.0046	0.0040	0.0045	0.0009
Data 3	AVG	0.0655	0.0524	0.0527	0.0558	0.0498	**0.0432**
	STD	0.0196	0.0010	0.0085	0.0062	0.0068	0.0049
Data 4	AVG	0.0326	0.0170	0.0246	0.0237	0.0226	**0.0149**
	STD	0.0293	0.0010	0.0122	0.0116	0.0120	0.0006
Data 5	AVG	0.0388	0.0374	0.0381	0.0377	0.0344	**0.0318**
	STD	0.0020	0.0011	0.0032	0.0011	0.0019	0.0034
Ranking	F-Test	5.2	3.4	4.2	5	2.2	1

Table 7. Comparison of VBSO with other well-known optimizers in terms of average of computation time

Benchmark	Measure	BGSA	BALO	BBA	BSSA	GA	VBSO
Data 1	AVG	512.75	725.03	441.42	857.84	**359.14**	481.83
	STD	13.5515	47.9891	25.9920	29.5709	21.9992	10.6841
Data 2	AVG	496.80	697.20	450.40	839.27	**373.75**	483.20
	STD	10.5753	37.7738	15.1373	21.6266	25.0813	8.1353
Data 3	AVG	500.09	743.93	449.04	839.09	**364.18**	482.94
	STD	7.9613	39.5577	18.1382	16.7807	39.0445	10.7459
Data 4	AVG	503.03	678.69	447.34	843.45	**370.99**	487.46
	STD	10.4878	40.8230	25.7421	24.1787	28.1989	13.1576
Data 5	AVG	494.96	696.57	455.66	870.90	**375.40**	473.71
	STD	9.0056	39.7433	36.5189	26.8016	27.3678	9.8778
Ranking	F-Test	4	5	2	6	**1**	3

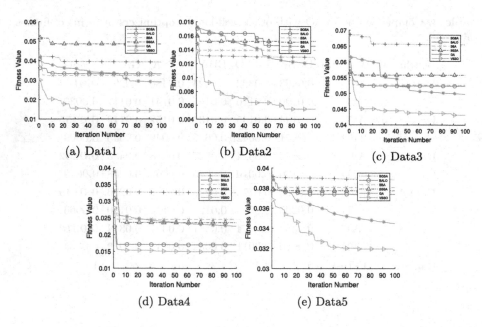

(a) Data1   (b) Data2   (c) Data3

(d) Data4   (e) Data5

Fig. 1. Convergence curves of compared algorithms

5   Conclusion and Future Directions

This paper proposed an effective FS approach based on the new Snake Optimizer for enhancing intrusion detection systems. Two variants of FS were developed, and the performance of the best variant, which is based on the V-shaped transfer function, was compared against several optimization algorithms to prove its efficiency in boosting IDSs. Five datasets that represent real IoT traffic were employed to examine the performance of the proposed VBSO approach purpose. Experimental results revealed that SO based on V-shaped transfer is superior to the S-shaped transfer function and outperforms other optimizers. Furthermore, VBSO obtained the best averages of accuracy and convergence rates over all considered datasets. Therefore, it can be said that the VBSO FS approach can be efficiently employed in IoT to boost intrusion detection systems.

SO is still new and its operations can be expanded in a number of new directions to handle more real-world datasets. One of the possible options is to integrate SO with hybrid wrapper-filter techniques and evolutionary-based FS systems. We will create new, improved binary SO variations to solve other complex binary optimization problems in the upcoming studies.

Acknowledgements. The research leading to these results has received funding from the Research Council (TRC) of the Sultanate of Oman under the Block Funding Program with agreement no. TRC/BFP/ASU/01/2019.

References

1. Conti, M., Dehghantanha, A., Franke, K., Watson, S.: Internet of Things Security and Forensics: Challenges and Opportunities. Elsevier, Amsterdam (2018)
2. Mafarja, M., Heidari, A., Habib, M., Faris, H., Thaher, T., Aljarah, I.: Augmented whale feature selection for IoT attacks: structure, analysis and applications. Future Gener. Comput. Syst. **112**, 18–40 (2020)
3. Hung, M.: GARTNER Inc Gartner Insights on How to Lead in a Connected World. GARTNER Inc. (2017)
4. Mahmoud, R., Yousuf, T., Aloul, F., Zualkernan, I.: Internet of things (IoT) security: current status, challenges and prospective measures. In: 2015 10th International Conference for Internet Technology and Secured Transactions (ICITST), pp. 336–341. IEEE (2015)
5. Hemdan, E.E.-D., Manjaiah, D.H.: Cybercrimes investigation and intrusion detection in internet of things based on data science methods. In: Sangaiah, A.K., Thangavelu, A., Meenakshi Sundaram, V. (eds.) Cognitive Computing for Big Data Systems Over IoT. LNDECT, vol. 14, pp. 39–62. Springer, Cham (2018). https://doi.org/10.1007/978-3-319-70688-7_2
6. Prabavathy, S., Sundarakantham, K., Shalinie, S.M.: Design of cognitive fog computing for intrusion detection in Internet of Things. J. Commun. Netw. **20**(3), 291–298 (2018)
7. Khan, M.A., Salah, K.: IoT security: review, blockchain solutions, and open challenges. Futur. Gener. Comput. Syst. **82**, 395–411 (2018)
8. Elrawy, M.F., Awad, A.I., Hamed, H.F.: Intrusion detection systems for IoT-based smart environments: a survey. J. Cloud Comput. **7**(1), 21 (2018)
9. Tan, P.-N., Steinbach, M., Kumar, V.: Introduction to Data Mining, 1st edn. Pearson Addison Wesley, Boston (2005)
10. Xue, B., Zhang, M., Browne, W.N.: Particle swarm optimization for feature selection in classification: a multi-objective approach. IEEE Trans. Cybern. **43**(6), 1656–1671 (2013)

11. Qiao, W., Tian, W., Tian, Y., Yang, Q., Wang, Y., Zhang, J.: The forecasting of PM2.5 using a hybrid model based on wavelet transform and an improved deep learning algorithm. IEEE Access **7**, 142814–142825 (2019)
12. Mafarja, M., Mirjalili, S.: Whale optimization approaches for wrapper feature selection. Appl. Soft Comput. **62**, 441–453 (2018)
13. Thaher, T., Chantar, H., Too, J., Mafarja, M., Turabieh, H.T., Houssein, E.H.: Boolean particle swarm optimization with various evolutionary population dynamics approaches for feature selection problems. Expert Syst. Appl. **195**, 116550 (2022)
14. Thaher, T., Heidari, A.A., Mafarja, M., Dong, J.S., Mirjalili, S.: Binary Harris Hawks optimizer for high-dimensional, low sample size feature selection. In: Mirjalili, S., Faris, H., Aljarah, I. (eds.) Evolutionary Machine Learning Techniques. AIS, pp. 251–272. Springer, Singapore (2020). https://doi.org/10.1007/978-981-32-9990-0_12
15. Chen, H., Heidari, A.A., Zhao, X., Zhang, L., Chen, H.: Advanced orthogonal learning-driven multi-swarm sine cosine optimization: framework and case studies. Expert Syst. Appl. **144**, 113113 (2020)
16. Xu, Y., et al.: An efficient chaotic mutative moth-flame-inspired optimizer for global optimization tasks. Expert Syst. Appl. **129**, 135–155 (2019)
17. Hashim, F.A., Hussien, A.G.: Snake optimizer: a novel meta-heuristic optimization algorithm. Knowl.-Based Syst. **242**, 108320 (2022)
18. Agrawal, P., Abutarboush, H., Talari, G., Wagdy, A.: Metaheuristic algorithms on feature selection: a survey of one decade of research (2009–2019). IEEE Access **9**, 26766–26791 (2021)
19. Hamamoto, A.H., Carvalho, L.F., Sampaio, L.D.H., Abrão, T., Proença Jr., M.L.: Network anomaly detection system using genetic algorithm and fuzzy logic. Expert Syst. Appl. **92**, 390–402 (2018)
20. Hajimirzaei, B., Navimipour, N.J.: Intrusion detection for cloud computing using neural networks and artificial bee colony optimization algorithm. ICT Express **5**(1), 56–59 (2019)
21. Selvakumar, B., Muneeswaran, K.: Firefly algorithm based feature selection for network intrusion detection. Comput. Secur. **81**, 148–155 (2019)
22. Ali, M.H., Al Mohammed, B.A.D., Ismail, A., Zolkipli, M.F.: A new intrusion detection system based on fast learning network and particle swarm optimization. IEEE Access **6**, 20255–20261 (2018)
23. Hodo, E., et al.: Threat analysis of IoT networks using artificial neural network intrusion detection system. In: 2016 International Symposium on Networks, Computers and Communications (ISNCC), pp. 1–6. IEEE (2016)
24. Panigrahi, A., Patra, M.R.: A layered approach to network intrusion detection using rule learning classifiers with nature-inspired feature selection. In: Pattnaik, P.K., Rautaray, S.S., Das, H., Nayak, J. (eds.) Progress in Computing, Analytics and Networking. AISC, vol. 710, pp. 215–223. Springer, Singapore (2018). https://doi.org/10.1007/978-981-10-7871-2_21
25. Su, J., He, S., Wu, Y.: Features selection and prediction for IoT attacks. High-Confidence Comput. **2**(2), 100047 (2022)
26. Diro, A.A., Chilamkurti, N.: Distributed attack detection scheme using deep learning approach for internet of things. Futur. Gener. Comput. Syst. **82**, 761–768 (2018)
27. Anthi, E., Williams, L., Burnap, P.: Pulse: an adaptive intrusion detection for the internet of things. In: Living in the Internet of Things: Cybersecurity of the IoT - 2018, pp. 1–4 (2018)

28. Mirjalili, S., Lewis, A.: S-shaped versus V-shaped transfer functions for binary particle swarm optimization. Swarm Evol. Comput. **9**, 1–14 (2013)
29. Li, J., Zhao, Z., Li, R., Zhang, H., Zhang, T.: AI-based two-stage intrusion detection for software defined IoT networks. IEEE Internet Things J. **6**(2), 2093–2102 (2018)
30. Kennedy, J., Eberhart, R. C.: A discrete binary version of the particle swarm algorithm. In: 1997 IEEE International Conference on Systems, Man, and Cybernetics. Computational Cybernetics and Simulation, vol. 5, pp. 4104–4108. IEEE (1997)
31. Rashedi, E., Nezamabadi-Pour, H., Saryazdi, S.: BGSA: binary gravitational search algorithm. Nat. Comput. **9**(3), 727–745 (2010)

Predicting Patient's Waiting Times in Emergency Department: A Retrospective Study in the CHIC Hospital Since 2019

Nadhem Ben Ameur[1,4], Imene Lahyani[1,2(✉)], Rafika Thabet[3,4],
Imen Megdiche[4,5], Jean-christophe Steinbach[6], and Elyes Lamine[3,4]

[1] National School of Engineers of Sfax, University of Sfax, Sfax, Tunisia
{nadhem.benameur,imen.lahyani}@enis.tn
[2] ReDCAD Laboratory, Sfax, Tunisia
[3] Centre Génie Industriel, IMT Mines Albi, Toulouse University, Albi, France
{rafika.thabet,elyes.lamine}@mines-albi.fr
[4] Institut National Universitaire Champollion, ISIS, Toulouse University,
Castres, France
imen.megdiche@univ-jfc.fr
[5] IRIT, Institut de Recherche en Informatique de Toulouse, Toulouse University,
Toulouse, France
imen.megdiche@irit.fr
[6] Centre Hospitalier Intercommunal Castres-Mazamet, Castres, France
jean-christophe.steinbach@chic-cm.fr

Abstract. Predicting patient waiting times in public emergency department rooms (EDs) has relied on inaccurate rolling average or median estimators. This inefficiency negatively affects EDs resources and staff management and causes patient dissatisfaction and adverse outcomes. This paper proposes a data science-oriented method to analyze real retrospective data. Using different error metrics, we applied various Machine Learning (ML) and Deep learning (DL) techniques to predict patient waiting times, including RF, Lasso, Huber regressor, SVR, and DNN. We examined data on 88,166 patients' arrivals at the ED of the Intercommunal Hospital Center of Castres-Mazamet (CHIC). The results show that the DNN algorithm has the best predictive capability among other models. By precise and real-time prediction of patient waiting times, EDs can optimize their activities and improve the quality of services offered to patients.

Keywords: Patient waiting times · Emergency department · Retrospective analysis · Data analysis · Information system

1 Introduction

Emergency Department Overcrowding (EDW) is a long standing issue affecting all healthcare systems. This is mainly caused by the imbalance between supply and demand for emergency services, which is becoming overwhelming due to the ageing population and the lack of human resources. ED overcrowding implies different consequences such as waiting times, delayed treatment or ambulance diversion. Prolonged waiting times may reduce the quality of care and increase the likelihood of adverse outcomes for patients with serious illnesses. Patient satisfaction is also affected, and more patients leave before being visited by a physician. Thus, waiting time is a key metric for measuring ED efficiency.

Measuring waiting time is a key challenge that was addressed by several works [3,5,6]. In [3], they test different machine learning techniques using predictive analysis applied to two large datasets from two emergency departments in medium-sized public hospitals located in average-populated areas of north-central Italy. [5] develop deep learning models alongside or instead of queuing theory techniques to predict waiting time in a queue using real low patient acuity data collected from electronic medical records in an ED in Saudi Arabia. [6] propose to use statistical learning algorithms to significantly improve waiting time predictions using routinely collected administrative data from a large Australian public tertiary health care centre.

In this paper, we propose a retrospective study of waiting times in the Emergency Department of a French hospital of Castres Mazamet CHIC. In our approach, we experiment with different machine learning techniques on a large amount of real data covering patient characteristics and emergency characteristics. We also evaluate the sliding window technique to improve prediction accuracy due to quality issues.

The particularity of our work is a specification of waiting time prediction techniques according to the organization of this hospital and for real data suffering from some quality problems.

Our aim is to identify the learning technique that provides the most accurate and real-time previsions of waiting times for patients arriving to the ED. We evaluate the accuracy of Lasso, Random Forest, Support Vector Regression, Artificial Neural Network using two forecasting error measures: the mean squared error and the mean absolute error.

The paper is structured as follows: in Sect. 2 we introduce base study concepts. Then in Sect. 3, we address some work in the literature. In Sect. 4, we describe our methodology. We present the results in Sect. 5. Section 6 summarises the study and the findings and identifies future directions.

2   Preliminaries

In this research, we design predictive models for patients' waiting times in the ED of the CHIC Hospital. The waiting time was defined in collaboration with the chief of Emergency Department and the chief of the Information System of the hospital. The waiting time is declined according to the existing waiting rooms: we have internal and external waiting times as described in Fig. 1. At CHIC ED, the waiting room is adjacent to the ED entrance, and there are no barriers for patients to exit the ED once their wait time exceeds their willingness to wait. In

46   N. B. Ameur et al.

Fig. 1, we describe the simple ED patient flow. Upon arrival, The patient goes to one of the three registration counters, where further personal information (e.g., address, occupation) are collected, and the patient arrival record registration is created. After the registration, the patient may experience a queue to see a triage nurse who records some basic identification details, presents a complaint, describes the nature of the visit, and assigns a triage level and a treatment box. After the triage, the patient proceeds to the designated waiting area if not called in for treatment immediately. A patient must wait in a designated waiting area until being called in to see a physician or an intern, or a treating nurse may come out to the waiting area and take vital signs and periodically check on the patient status while waiting to be called in for treatment. Different ED teams generally treat patients in different ED areas. For example, ED also operates a Short-Term Hospitalization Unit (STHU) specifically designed for short-term treatment (often completed within 24 h). This inpatient unit under ED management provides health services to patients initially assessed, triaged, and treated in the ED. Most often, patients with selected medical conditions that require further observation, assessment, and reassessment but are unlikely to require a prolonged admission to the hospital are assigned to STHU. These patients are not considered by this study.

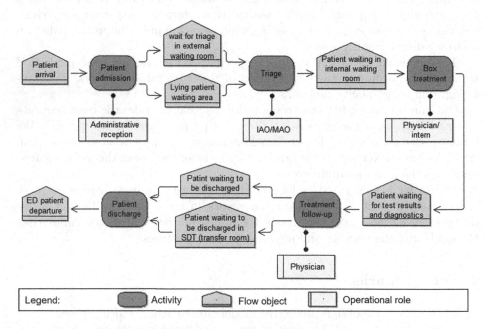

Fig. 1. Business process model of ED patient flow

3   Related Work

This section provides a comprehensive review of the works related to waiting time prediction in public emergency department rooms (EDs). Previous studies have demonstrated that prolonged waiting periods cause patients' frustration, anxiety and dissatisfaction [10]. In addition, patients are concerned about the quality of the information provided in EDs and how to interpret it [4]. Indeed, the current ED waiting times reporting systems are limited and generally based on rolling average or median estimators. Waiting time is one of the most important quality indicators for emergency department [8], several studies proposed various approaches to optimize patients waiting times prediction and enhance EDs resources management efficiency. [5] combined deep learning with queuing theory to predict waiting time in a queue using real low patient acuity data. They provided a guideline for queue waiting time analysis not only in healthcare but also in other sectors considering model understand and feature extraction process. [3] implemented a combination of approaches including machine learning and various categories of features to predict patient waiting time in real time. [2] used ordinal logistic regression and data mining techniques to develop models to classify ED patients in terms of their waiting and treatment times. Finally, [6] implemented quantile regression and statistical learning algorithms for a large set of predictors using comprehensive EMR from the ED information system. They developed models to predict ED wait time for patients with low acuity assigned to the ED waiting room and differentiated the low acuity patients into waiting room and non-waiting room patients for the purpose of waiting room prediction.

Our study differs from previous research on this topic as we predicted patients waiting times specific to CHIC ED and we implemented various ML and multi-DL optimization algorithms to improve accuracy. We used sliding window algorithm to detect outliers in data. In addition, we accounted for various predictors by extracting various categories of features such as patient's related features, queue-based and patient condition's severity score-based features. Our models with multiple ML and Multi-DL optimization algorithms can be applied in evaluating the wait times of patients with in the emergency department. Consequently, the models proposed in this study can be used to provide insights to medical staff in the emergency room to determine patient waiting time in the queue using EHR data.

4   Methodology

In this section, the extracted data and the used dataset are described, and our methodological process is detailed. As presented in Fig. 2, our methodology includes five main steps: data cleaning, feature engineering, dataset analysis, dataset preprocessing, Machine Learning models, and model evaluation. In the following, a detailed description of each step will be presented.

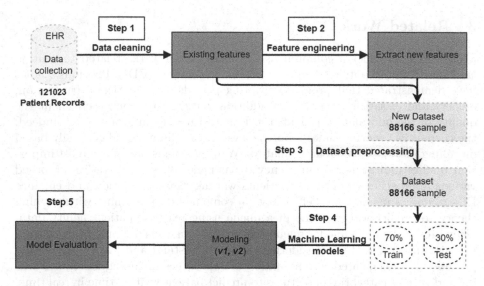

Fig. 2. Proposed methodology

4.1   Data Description

We used real data from the CHIC hospital ED. Electronic Health Records (EHR) tend to be complex and challenging to manage [9]. The Conceptual Data Models (CDM) of the databases used contains more than 1000 entity tables, from which 22 tables were selected to work on ED activities. We examined data on 121,023 patient's arrivals of the hospital ED spanning the period from January 2019 to April 2022. The extracted raw data contain patients-related information and their associated ED episodes, from the registration and triage process to ED discharge or hospital admission timestamps. In order to build a relevant set of features, we essentially proceeded in three ways: the identification of relevant features previously approved in the scientific work for waiting time prediction, the exchange with the staff of the emergency services and the staff of the IT department of the CHIC hospital, and finally, the use of the data collected by the information system (databases and reports). The raw data was then cleaned and used to extract useful features for predicting patient waiting times specific to the CHIC hospital context. The features included in the data are summarized in Table 1.

4.2   Data Cleaning

The extracted raw data has several quality issues. It contains duplicate, missing, irrelevant data and outliers. In order to solve these problems, we proceeded in 4 phases:

- **Duplicated samples removal:** The extracted raw data contains few redundant patient records (PR) since the data used is the combination of the

Table 1. Description of raw data features

Features	Type	Features	Type
Patient Record PR_id	Numerical	Age	Numerical
Sex	Categorical	Postal code	Categorical
ED patients count	Numerical	Patient arrival status	Categorical
Raison for admission	Categorical	Severity score	Categorical
Arrival entry mode	Categorical	Arrival exit mode	Categorical
Patient arrival ts	Datetime	Ts nurse triage	Datetime
Nurse location ts_start	Datetime	Nurse location ts_end	Datetime
Internal waiting ts_start	Datetime	Internal waiting ts_end	Datetime
Box ts_entry	Datetime	Box ts_quit	Datetime
Box type	Categorical	Pattern code	Categorical
Transfert room wait ts_start	Datetime	Transfert room wait ts_end	Datetime
PR discharge ts	Datetime	PR exit ts	Datetime
Patient exit ts	Datetime	STHU patient count	Numerical
Triage ts validity	Binary	Internal waiting time calculability	Binary
Internal waiting time deb ts	Datetime	Internal waiting time end ts	Datetime

extractions from different resources (ED database, hospital database). The library used to remove duplicates is Pandas, which contains various built-in data preprocessing methods. The discarded sample size is 344 samples.

– **Filter patients by age:** In this research, we found it relevant to work on predicting waiting times for adults. Firstly, because children always have priority over adults, so they don't wait very long. And secondly, children have separate waiting rooms and treatment boxes from adults, from which they are not drawn into the flow of adult patients. After filtering data by age (age ≥ 12), The discarded sample size is 18,879 samples.

– **Formatting dates:** Extracting data from multiple databases results in a heterogeneous dates format. Most of the extracted features listed in Table 1 are of type Datetime. Therefore, we performed a data formatting step to ensure that all datetime features have the same format (dd/MM/YYYY hh:mm).

– **Clean "Ts nurse triage" feature:** The data extraction period extends from January 2019 to April 2022 during the Covid-19 pandemic. This causes delays in the storage of patient records in the CHIC databases due to material or human matters. The feature "Ts nurse triage" (Table 1) is a special feature that corresponds to the timestamp of patient records creation in the ED database. As a result, there were some cases where the triage timestamp was severely delayed and calculating patient wait time was impossible. To reduce the number of discarded samples, we used the "Nurse location ts_start" feature (Table 1) because the triage process begins when the patient is in the nurse care area. Unfortunately, the "Nurse location ts_start" feature contains missing values, so it is not possible to recover all patient records with invalid triage timestamps. After the cleaning process, the discarded sample size is 280 samples.

– **Filter patients by ED waiting time predictability:** The predictability of patient waiting time depends on datetime features listed in Table 1. Because most of these features contain missing or invalid values, a cleanup process is required to ensure predictability of ED patient waiting time. We used the redundant features where there was timeline overlap in the patient path (e.g. the "Internal waiting ts_star" feature and the "Box ts_entry" feature must have the same values as the patient leaves the internal waiting room when he enters the assigned box for him)

4.3   Feature Engineering

The cleaned data was used to generate the dataset for modeling patient waiting time prediction algorithms. We considered a set of 30 features, as listed in Table 2. The patient-related features are already included in the raw data. It includes patient' demographic information such as age and sex, and information about the patient' status at arrival, the reasons for admission and information about nurse diagnosis (pattern code). To account for daily variations and highlight the seasonal effect, we investigated arrival time variables namely the time of day along with the day of the week, the week, the month, the quarter and a Boolean feature to identify weekend patient arrivals.

The higher the priority of new arrivals, the more significant their impact on the waiting time. Patients with high severity scores (score 0 is the most severe score, score 5 is patients with mental disorders) consume human and material resources for a long periods, which causes patients with the least severe scores to wait for longer periods. Thus, we developed six features corresponding to severity score used in CHIC ED. To calculate Condition severity score based (CS)-based features for each observation, we counted the number of patients based on their severity score in the admission timestamp.

Alongside the features commonly used in the healthcare literature, we also explored the use of queue-based features [3, 6], which report the current state of the ED, to improve prediction accuracy, as suggested Queue-based features capture the influence of new ED arrivals on the waiting time of patients within the ED. Queue-based features determine the crowding level of the ED system. They measure the queue of patients waiting at various stages, from registration to discharge (e.g., the queue in internal waiting room or for transfer to the short-term hospitalization units (STHU)). To derive internal queue-based features, we used different datetime features in Table 1 to count patient in different area in the admission timestamp for each observation.

Following [3], we incorporated the rolling average of the waiting times of the last 15 patients who presented to the ED as an additional candidate feature, as well as the bed occupancy (the ratio of occupied bed in the hospital) and the waiting rooms occupancy (the ratio of patients in waiting rooms by the total number of patients in the ED).

Table 2. Feature description

Categories	Features
Patient-related (P)	Age
	Sex
	Residence
	Arrival mode
	Patient status at entry
	Reason for admission
	Triage score
	Pattern code
Related to time of arrival (TA)	Hour of day
	Day of week
	Week
	Month
	Quarter
	Is_weekend
Condition severity score based (CS)	Number of patients score_0
	Number of patients score_1
	Number of patients score_2
	Number of patients score_3
	Number of patients score_4
	Number of patients score_5
Queue-based (Q)	Patients in external waiting room
	Patients in internal waiting room
	Treated patients waiting for discharge
	Discharged patients waiting for exit
	Patients in short-term hospitalization units (STHU)
	STHU_saturated
	ED patients count
ED and hospital indicators (I)	Bed occupancy
	Waiting rooms occupancy
	Wait times rolling average

All the process of feature engineering requires implementing algorithms for each category of features. Due to space limitation, we present only Algorithm 1 that calculate the patient number in the external waiting room.

Algorithm 1: Patient count in external waiting room algorithm

---

VARIABLES :   *df_slice* : DATAFRAME
                   df_sorted : DATAFRAME
                   ranges : DICTIONNARY
                   range_min, range_max, row_index : INTEGER
                   other_patient_arrival_ts, other_patient_TS_nurse_triage,
patient_arrival_ts : DATETIME
                   PR_id, other_patient_PR_id : INTEGER
                   counter, index : INTEGER
INPUT   : *df_sorted, PR_id, patient_arrival_ts, ranges*
OUTPUT   : *counter, index*

begin
    counter ← 0
    range_min ← ranges["min"]
    range_max ← ranges["max"]
    index ← range_min
    df_slice ← df_sorted.iloc[range_min : range_max]
    FOR *row_index* **in** *range (range_min, range_max)* **do**
      other_patient_arrival_ts ← df_slice.loc[row_index,
    "Patient_arrival_ts"]
      other_patient_TS_nurse_triage ← df_slice.loc[row_index,
    "Ts_nurse_triage"]
      other_patient_PR_id ← df_slice.loc[row_index, "PR_id"]
      IF *patient_arrival_ts* < *other_patient_arrival_ts* **do**
        Break
      IF *other_patient_PR_id* ≠ *PR_id* **and**
    other_patient_arrival_ts ≤ *patient_arrival_ts* ≤
    other_patient_TS_nurse_triage **do**
        counter ← *counter + 1*
    IF *counter* ≡ 1 **do**
      index ← *row_index*
    Return *counter, index*
end

The new dataset included 88,166 observations and 30 variables. The variables included in the dataset are summarized in Table 2.

We implement other feature engineering scripts and make our code available on GitHub.[1]

4.4 Dataset Preprocessing

Data preprocessing is considered as the main step in any ML classification process. It refers to the transformations applied to data before feeding it into our ML algorithms. The goal is to convert the raw data into a clean dataset. We had to apply a sliding window technique and normalization to deal with the presence of outliers in the used numerical features. As for missing values issue, we

---

[1] https://github.com/nadhembenameur99/CHIC-ED-patients-waiting-times-predicti on.git.

applied an imputing technique. Data encoding is applied to categorical features to convert them to numerical format, as ML algorithms require numerical data.

- **Sliding window concept for outlier detection** [1]: Firstly, the sliding window size is set to h and the data size is set to N. Then a sample formed by data from i to i + h, where $1 \leq i \leq N - h$, is used to estimate the upper and the lower bands of values inside the window. The window slide by one, the sample is updated, and the new upper and lower bands are produced.
- **Missing values imputation:** For imputing missing data, the most frequent imputation method is used to impute "Residence" and "Pattern code" features. In this approach, we impute each missing value by the most frequent value for each feature.
- **Data normalization:** The goal of normalization is to transform features to be on a similar scale. This improves the performance and training stability of the model. This transformation converts all the features into the same scale and harmonizes the data structure.
- **Categorical data encoding:** For encoding categorical data, the frequency encoding method is used to encode "Residence" and "Pattern code" features. In this approach, we simply encode each unique value by its frequency. This method is used to ignore the natural ordered relationship between integer values and to give more importance to the most frequent values in patient waiting time prediction. As for the other categorical features, they are already in numerical mode, so there were no need to use an encoder.

4.5 Machine Learning Models

The next step is to develop learning regression algorithms. The objective of this study is, therefore, to apply more preprocessing methods and then to compare supervised techniques based on linear regression (Huber, Lasso and SVR) which are sensitive to outliers, with a decision tree' based ensemble method (RF) and a deep neural network' based method. These algorithms are widely used [3,5,6] because they flexibly explore connections in the data, identify relevant features, are compatible with large datasets, and prevent overfitting [6]. As it is a supervised approach, we split the initial dataset into a training dataset (70%) and a test dataset (30%). To emphasize the relevance of the features, we built two versions of prediction models (V1 and V2) according to each feature category (Table 2) to be included. As for the sets of features to be included in the different versions of the predicting model, they are defined as follows:

1. Models V1: This includes Patient-related (**P**), Related to time of arrival (**TA**), Condition severity score based (**CS**) and ED and hospital indicators (**I**) categories for training the prediction algorithms (P + TA + CS + I).
2. Models V2: This version includes features used for training models V1 plus the Queue-based (**Q**) features (P + TA + CS + I + Q).

We implemented the selected learning techniques by using Anaconda distribution of Python programming languages for data science, Tensorflow 2, Keras and scikit-learn libraries.

To further improve our training model, parameter tuning is conducted using the Grid Search for hyperparameter optimization. For each algorithm, we chose to tune the most important parameters. In the case of SVR, the type of kernel and its degree are adjusted. For RF, the number of trees in the forest is used in addition to the maximum depth of the trees. Finally, when using Lasso, we tuned the alpha constant, controlling the L1 regularization strength.

- **Random Forest (RF)** [7]: RF is a learning technique that generates and combines binary decision trees, while also aggregating the results. Decision trees are constructed by using a bootstrap sample of the training data and randomly choosing a subset of features at each node [3]. As for the fine-tuning, in our case we set the minimum number of samples required to be at a leaf node **min-samples-leaf = 3**, the number of trees in the forest **n-estimators = 40** and the rest as default.
- **Lasso:** Lasso is a regularized linear regression method that can be used to select significant parameters from features. Lasso has the advantage of reducing the number of features in the model. So if a dataset has many features, Lasso can identify and extract the most important ones. However, when a set of features is highly correlated, Lasso tends to randomly choose one of the multicollinear features, regardless of context. For the fine-tuning, in our case we set the alpha constant **alpha = 0.02** and the rest as default.
- **Huber Regressor:** The Huber regressor is a linear ML regression technique that is robust to outliers. The idea is to use a different loss function than traditional least squares. In statistics, the Huber loss is a specific loss function that is often used for situations of robust regression problems where outliers are present that can affect the performance and accuracy of the least-squares-based regression. For the fine-tuning, in our case we used the default parameters.
- **Support Vector Regression:** SVR belongs to the family of generalized linear models that aim to make a predictive decision based on a linear combination of features derived from features. SVR provides the flexibility to define how much error is acceptable in the model and finds an appropriate line (or hyperplane in higher dimensions) to fit the data. For SVR fine-tuning, in our case we set the kernel type to polynomial **kernel = 'poly'** and the rest as default.
- **Multi-Layer Perceptron (MLP):** A multi-layer perceptron (MLP) is a feed-forward artificial neural network that generates a set of outputs from inputs. It is characterized by multiple layers of input nodes connected as a directed graph between the input and output layers. As for the network architecture, we set two hidden layers, 25 neurons for the first hidden layer, 18 neurons in the next hidden layer, and one neuron in the output layer.

5 Experimental Results

In this section, the experimental results and model evaluation are presented. Python programming language is used in the implementation phase and a PC

Table 3. Results of the models performance evaluation

	Algorithms	Rolling average	LR	RF	Lasso	Huber	SVR	MLP (Adam)	MLP (Rmsprop)
Model V1	MAE	48.52	43.49	38.97	43.58	41.86	40.62	37.62	**37.6**
	RMSE	75.53	57.85	**53.92**	57.93	58.54	59.48	55.86	56.14
Model V2	MAE	48.52	43.03	38.86	43.1	41.51	39.97	**37.18**	37.29
	RMSE	75.53	57.67	**53.73**	57.75	58.66	58.89	56.01	55.75

with an Intel (R) Core (TM) i7-10870H processor with 16 GB of RAM. Regression metrics such as MAE and RMSE are used to evaluate the performance of the system. These metrics, most commonly used in the literature, are useful for comparing model performance. MAE represents the absolute difference between the predicted and observed values and its robustness to outliers in the dataset. The RMSE is widely used because of its theoretical significance in statistical modelling [3]. On the other hand, RMSE is sensitive to outliers and for this reason we decided to use MAE with the RMSE.

The results of all the metrics used for the models evaluation (MAE, RMSE) are presented in Table 3 for both model versions (V1 and V2).

Analyzing the results from Table 3 regarding the performance of the models, the model MLP (Adam) outperformed all the other with the lowest value of MAE equal to 37.18 and RMSE equal to 56.01. RF is a method based on Decision Tree and apply bagging algorithms, we noticed that the bagging method is suitable to the dataset used with many outliers, as it has the lowest value of RMSE equal to 53.73.

Concerning the results of ML models applied on dataset with Queue-based features included (models V2), the results have improved but they are very close to those of models V1 comparing the evaluation metrics used.

Figures 3 and 4 show the actual and the predicted patient waiting time for models V1 and V2, respectively.

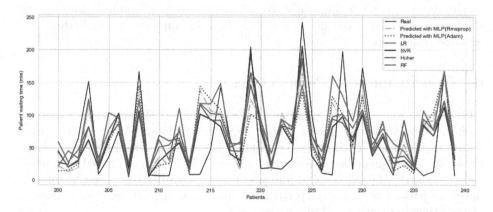

Fig. 3. Waiting time predicted vs. actual for Models V1

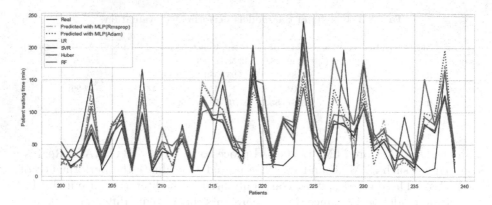

Fig. 4. Waiting time predicted vs. actual for Models V2

The behaviour of the prediction curves is almost the same for both model versions. Models V2 are slightly better in predicting waiting time values that tend to be in the range of outliers. This confirms that Queue-based features can enhance models performance, and that the proposed set of features can be extended further with other categories of features, in order to reduce the prediction error threshold. The high variance in the target feature (waiting time) prevent models from effectively train and causes overfitting. The trained models tend to underestimate the prediction when compared to the real values. RF and MLP models in V1 and V2 are capable of following data trends and learn better than other regression models. This is a promising result in a complex and dynamic context like an ED.

From the experiments, MLP (Adam) outperformed RF, Lasso, Huber regressor and SVR in the MAE metrics whereas RF outperformed Lasso, Huber regressor, SVR MLP (Adam) and MLP (Rmsprop) models in the RMSE metrics. For a real time case study, MLP (Adam) and RF are a good algorithms regarding to the features used in this study. We had to find a good compromise between the two constraints in our design which are a minimum error in predictions and less execution time.

6   Conclusion and Perspectives

This paper proposed a comparative study between ML/DL prediction models, namely Random Forest, Lasso, Huber regressor, SVR and MLP models (with Adam and Rmsprop optimizer) to evaluate ED patient's waiting times using real data. We used queue-based features combined with patient-related, condition severity-based, ED and hospital indicators, and time of arrival based features to optimize patient waiting time prediction along with ML/DL techniques instead of traditional approaches (Rolling average).

The study contributes to the current literature as an extensive and structured comparative analysis that establishes the real value of different learning techniques in complex and dynamic environments like EDs. We have demonstrated the accuracy and suitability of predictive models in a real application context. This study also offers a practical contribution through the use of real data from an ED.

The results showed that the MLP model achieved better prediction accuracy (based on the MAE metric) than the traditional approaches. Future and extended work of this research are as follows: More information from ED could be implemented into the model, like parameters for medical staff schedules. In addition, different datasets from other hospitals and sites could be implemented. Furthermore, various machine learning algorithms could be further applied, including linear and nonlinear regression like the XGBoost ML algorithm. Finally, meteorological and event-related features could be used to extend the feature set and integrate environmental effects.

References

1. Amor, L.B., Lahyani, I., Jmaiel, M.: Recursive and rolling windows for medical time series forecasting: a comparative study. In: 2016 IEEE Intl Conference on Computational Science and Engineering, CSE 2016, and IEEE Intl Conference on Embedded and Ubiquitous Computing, EUC 2016, and 15th Intl Symposium on Distributed Computing and Applications for Business Engineering, DCABES 2016, Paris, France, 24–26 August 2016, pp. 106–113. IEEE Computer Society (2016). https://doi.org/10.1109/CSE-EUC-DCABES.2016.169
2. Ataman, M.G., Sarıyer, G.: Predicting waiting and treatment times in emergency departments using ordinal logistic regression models. Am. J. Emerg. Med. **46**, 45–50 (2021)
3. Benevento, E., Aloini, D., Squicciarini, N.: Towards a real-time prediction of waiting times in emergency departments: a comparative analysis of machine learning techniques. Int. J. Forecast. (2021)
4. Hemaya, S.A., Locker, T.E.: How accurate are predicted waiting times, determined upon a patient's arrival in the emergency department? Emerg. Med. J. **29**(4), 316–318 (2012)
5. Hijry, H., Olawoyin, R.: Predicting patient waiting time in the queue system using deep learning algorithms in the emergency room. Int. J. Ind. Eng. **3**(1) (2021)
6. Pak, A., Gannon, B., Staib, A., et al.: Forecasting waiting time to treatment for emergency department patients. OSF Preprints **26** (2020)
7. Patil, P., Thakur, S.D., Kasap, N.: Patient waiting time prediction in hospital queuing system using improved random forest in big data. In: 2019 International Conference on Issues and Challenges in Intelligent Computing Techniques (ICICT) (2019)
8. Soremekun, O.A., Takayesu, J.K., Bohan, S.J.: Framework for analyzing wait times and other factors that impact patient satisfaction in the emergency department. J. Emerg. Med. **41**(6), 686–692 (2011)
9. Tomar, D., Agarwal, S.: A survey on data mining approaches for healthcare. Int. J. Bio-Sci. Bio-Technol. **5**(5), 241–266 (2013)
10. Ward, P.R., et al.: 'waiting for' and 'waiting in' public and private hospitals: a qualitative study of patient trust in south Australia. BMC Health Serv. Res. **17**(1), 1–11 (2017)

Natural Language Processing

Topic Modeling on Arabic Language Dataset: Comparative Study

Aly Abdelrazek[1]([✉])(iD), Walaa Medhat[1,2](iD), Eman Gawish[1](iD),
and Ahmed Hassan[1,3](iD)

[1] Nile University, El Sheikh Zayed, Giza 16453, Egypt
al.abdelrazek@nu.edu.eg
[2] Benha University, Al-Qalyubia, Egypt
[3] Ain Shams University, Cairo, Egypt

Abstract. Topic modeling automatically infers the hidden themes in a collection of documents. There are several developed techniques for topic modeling, which are broadly categorized into Algebraic, Probabilistic and Neural. In this paper, we use an Arabic dataset to experiment and compare six models (LDA, NMF, CTM, ETM, and two Bertopic variants). The comparison used evaluation metrics of topic coherence, diversity, and computational cost. The results show that among all the presented models, the neural BERTopic model with Roberta-based sentence transformer achieved the highest coherence score (0.1147), which is 36% above Bertopic with Arabert (the second best in coherence). At the same time, the topic diversity is 6% lower than the CTM model (the second best in diversity) at the cost of doubling the computation time.

Keywords: Topic models · BERTopic · Arabic

1 Introduction

Topic models (TMs) have been used successfully in mining large text corpora where a topic model takes a collection of documents as input and then attempts, without any supervision, to uncover the underlying topics in this collection [8]. Each topic describes a human-interpretable semantic concept. TMs then provide a latent interpretable representation of documents according to the conceived topics. This representation can be used for other downstream tasks, such as text classification or aspect-based sentiment analysis.

Since the introduction of Latent Semantic Analysis (LSA) in 1990 by [10], several techniques for topic modeling have been developed. They make different assumptions about the corpus, the documents' representation, and the topics. Three major model groups are recognized: algebraic, probabilistic, and neural. Algebraic models were developed first during the 1990s [10]. In 2003, Bayesian Probabilistic Topic Models (BPTMs) gained popularity with the development of [8]. BPTMs are computationally efficient owing to their small number of parameters. BPTMs are also simple to deploy, easy to interpret, and can be

© The Author(s), under exclusive license to Springer Nature Switzerland AG 2022
P. Fournier-Viger et al. (Eds.): MEDI 2022, CCIS 1751, pp. 61–71, 2022.
https://doi.org/10.1007/978-3-031-23119-3_5

extended into more complex models. Accordingly, they dominated the research venues till 2015. More neural elements are deployed in TMs after 2015 [9], such as [11]. Some Neural Topic Models (NTMs) use contextualized representation to represent the documents instead of the usual bag of words [6,13]. These models can use pre-trained models on large corpora, such as BERT variants. NTMs are flexible and scalable. Over the past several years, NTMs have proven to perform better in many cases than previous models.

Most of the developed techniques for topic modeling are language agnostic. The models can train on any vocabulary. However, once trained, they can be used only with documents having the same fixed vocabulary specific to the training. A trained model cannot handle unknown tokens and cannot be easily applied to other languages. Moreover, in NTMs deploying various embeddings to represent the input corpus, these models' performance depends on the quality of the obtained embeddings specific to the training language. This language dependence creates different challenges for different languages that dictate different handling.

Arabic is challenging for many reasons. It is an exceptionally morphological-rich language, and it is also highly derivational and inflectional. Moreover, there is an inherent ambiguity in the Arabic words with many polysemous and synonymous words. This ambiguity adds to the complexity of the Arabic language topic models. Furthermore, Arabic has three main varieties: classical Arabic (CA), Modern Standard Arabic (MSA), and Arabic Dialect (AD).

This paper evaluated various algebraic, probabilistic, and neural topic models. LDA and NMF were used to set the scores benchmark and then three different state-of-the-art neural topic models were used, Contextualized Topic Model (CTM) [7], Embedded Topic Model (ETM) [11], and BERTopic [12] which uses Bidirectional Encoder Representations from Transformers (BERT). With BERTopic, two transformers were used. The first is Arabert, an Arabic pre-trained language model based on Google's BERT architecture which was used to obtain document embeddings. The second model is "symanto/sn-xlm-roberta-base-snli-mnli-anli-xnli" which is a multilingual Siamese network model trained on four datasets for zero-shot and few-shot text classification tasks. It is based on xlm-roberta-base. This model is used to obtain sentence embeddings and is well-suited for sentence similarity tasks.

The results show that BERTopic with the "symanto/sn-xlm-roberta-base-snli-mnli-anli-xn" transformer model achieved the best topic coherence score. This model also achieved the second best topics diversity score. These results required high computational resources, however.

The main contribution of this paper is that five topic models of various types are presented and compared for the Arabic topic modeling task. These five models include three state-of-the-art recently published models. Moreover, two transformers were used to obtain embeddings for BERTopic to emphasize the impact of embedding model selection on the results. We took a holistic approach to evaluate the models, which gives a clearer picture of the models' performances. The models' performance was thus evaluated against a set of metrics: coherence, diversity, and computational efficiency. The results are interesting since no single model is best across all metrics. We also noted the effect of the dataset characteristics on the models' performance.

The rest of this paper is organized as follows: related work was presented in Sect. 2. Then, the conducted experiments were discussed in Sect. 3. The results were presented and discussed in Sect. 4. Finally, Sect. 5 concludes the work done and suggests future work.

2   Related Work

Some of the recent work on topic modeling for Arabic is surveyed in this section. In [17], the authors surveyed different modeling techniques. Their findings point to the dominance of LDA in the research venue. More recent work started incorporating LDA with other techniques, such as word2vec embeddings. In the context of Arabic social media content, [4] used NMF to extract seven topics from hate tweets during COVID-19 pandemic. In [5], the authors used fixed-length embeddings to represent Quran verses and then clustered the verses. The obtained clusters were then evaluated against an annotated corpus to verify the relationship between different verses of the Quran. In [1], the authors experimented with BERTopic and compared it to LDA and NMF. They considered the generated topics' coherence as a basis for comparison. While their work highlights the power of contextualized embeddings obtained from BERT language models, they did not consider other important evaluation metrics for the models, such as diversity and computational time. They also did not compare BERTopic against other recently developed neural models achieving state-of-the-art results such as ETM [11] and CTM [7]. In [3], the authors proposed a machine learning-based approach, using BERTopic, to improve cognitive distortions' classification of the Arabic content over Twitter. The authors of [2], however, applied LDA to classify topics on sustainability using Arabic text. Their work summarized subtopics that matter to various sustainability areas, demonstrating the power of topic modeling in capturing and analyzing various aspects of unstructured data.

3   Experimental Work

As typical in Natural Language Processing (NLP) pipelines (Fig. 1), we have several stages along the pipeline where each stage can have one from several configurations. We ran the experiments at different configuration settings and evaluated the results against the evaluation metrics.

3.1   Dataset

The first stage is dataset acquisition. We used the dataset for Arabic Classification (https://data.mendeley.com/datasets/v524p5dhpj/2). This dataset is

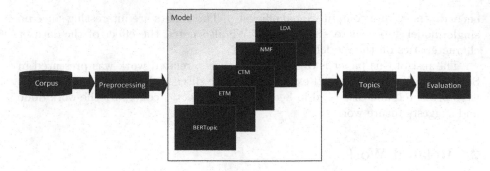

Fig. 1. System design

a collection of Arabic texts written in the modern Arabic language and collected from 3 Arabic online newspapers: Assabah, Hespress, and Akhbarona. The text contains alphabetic, numeric, and symbolic words. The dataset comprises 111,728 documents categorized into five classes: sport, politic, culture, economy, and diverse. Note that the dataset is clean and does not require heavy preprocessing. It is also recent and covers several topics, making it fit our use case. It is also publicly available and has been used previously in [1], which provides a basis for comparison with their work. In this study, we randomly picked 10K documents from this dataset to conduct our experiments. This size selection drastically decreases the required computational resources and provides a sound basis for experimentation with various models.

3.2 Data Preprocessing

The second stage is for preprocessing the datasets. Preprocessing includes tokenization, removing punctuation, removing stopwords, tagging, and constructing n-grams. Text normalization is essential. In this study, and for simplicity, only 1-gram tokens were included, and all but noun tokens were removed. This preprocessing results in a smaller number of tokens and slightly shorter documents which challenges the neural models whose performance is impacted by the corpus's vocabulary size and the length of the documents [19]. The preprocessing steps were implemented using CAMEL tools [15] (Table 1).

3.3 Model

The third stage is the actual model. The configurable items are the model itself and its hyper-parameters. Various topic models from different model families were used to compare models and model families. Latent Dirichlet Allocation (LDA) and Non-negative Matrix Factorization (NMF) were used first to set the scores benchmark, and then three different state-of-the-art neural topic models were used, Contextualized Topic Model (CTM), Embedded Topic Model (ETM),

Table 1. Dataset classes

Category	Count
Sports	4009
Politics	1813
Diverse	1617
Economy	1303
Culture	1258

and BERTopic, which uses Bidirectional Encoder Representations from Transformers (BERT). These models were chosen because they are the current state-of-the-art neural models that outperformed prior models and promise better topic quality. The OCTIS library [19] was used for the LDA, NMF, CTM and ETM models. For BERTopic, the open source library was used with two transformers available on the Hugging Face transformers library. The first is Arabert; an Arabic pre-trained language model based on Google's BERT architecture which was used to obtain document embeddings. The second model is "symanto/sn-xlm-roberta-base-snli-mnli-anli-xnli" which is a multilingual Siamese network model trained for zero-shot and few-shot text classification tasks. It is based on xlm-roberta- base model and is pre-trained on four datasets. This model is used to obtain sentence embeddings and is well-suited for sentence similarity tasks.

Typically, the post-processing includes the removal of stop-words after modeling, which is reported to provide similar results but is more transparent [18]. Also, merging topics into one and topics auto-labeling are familiar post-processing steps. In this work, no further post-processing is performed apart from the merging of similar topics by the BERTopic algorithm. This choice is made to ensure a fair comparison between the basic model variants without introducing other factors that could impact the results.

3.4 Latent Dirichlet Allocation (LDA)

LDA is a graphical probabilistic topic model. It is a bag of words model that represents a document as a dense vector. LDA theorizes a generation process as follows: each document i is generated, one word at a time, through sampling a topic from the distribution of topics for this document d, then sampling a word from this topic. The distribution of topics is drawn from a Dirichlet distribution where each component -topic- in the sampled mixture is independent of the other. This generative process defines a joint probability distribution over the observed documents and the hidden topic structure. LDA then attempts to infer the hidden topics from the observed words.

3.5 Non-negative Matrix Factorization (NMF)

Non-Negative Matrix Factorization is applied for topic modeling. In this context, the input is the Term Frequency - Inverse Document Frequency (TF-IDF) term-document matrix. This linear algebra technique then factorizes the input matrix into two non-negative matrices; word-topics matrix and topics-documents matrix.

3.6   Contextualized Topic Models (CTM)

CTM is a neural model that combines the power of contextualized representations with neural topic models [7]. Specifically, CTM extends neural ProdLDA (Product-of-Experts LDA): a neural topic model that implements black-box variational inference to include contextualized representations. In this study, CTM with AraBERT was used.

3.7   ETM

ETM defines words and topics in the same embedding space. In ETM, the likelihood of a word under ETM follows a categorical distribution whose natural parameter is given by the dot product between the word embedding and its assigned topic's embedding. ETM models documents by learning interpretable topics and word embeddings. It is usually robust to datasets with large vocabularies characterized by abundant rare and stop words.

3.8   BERT Models

The open-source BERTopic model was used with various pre-trained transformer language models to obtain embedding representations. The embeddings are then passed to BERTopic, which extracts the document embedding and then uses both the Uniform Manifold Approximation and Projection (UMAP) algorithm to reduce the embedding dimensionality and the Hierarchical Density-Based Spatial Clustering of Applications w/Noise (HDBSAN) algorithm for document clustering. BERTopic provides various options to extract document embeddings. One way is using the sentence-transformers package. Additionally, we can use any Hugging Face transformers model utilizing the Flair framework. In this study, two transformers were used; Arabic Arabert and another multilingual model named "symanto/sn-xlm-roberta-base-snli-mnli-anli-xnli". The number of topics was set at 5 for both transformers (Fig. 2).

Fig. 2. BERTopic model

3.9 Evaluation

A topic is a discrete distribution over words. This set of words is evaluated for being human-interpretable. For this to occur, the generated words should be associated with a single semantic meaning. One way of assessing the interpretability is to evaluate the coherence of the words in the generated topic [14].

In this work, we used the normalized pointwise mutual information, $NPMI$, between word pairs to calculate topics' coherence. The PMI between word pair (wi, wj) is calculated as below

$$PMI\ (w_i,\ w_j)\ =\ \frac{log\ p\ (w_i,\ w_j)}{p(w_i)p(\ w_j)} \tag{1}$$

PMI compares the probability of two words occurring together to what this probability would be if the words were independent. It can be normalized between $[-1, +1]$ resulting in -1 for words that are never occurring together, 0 for independence, and $+1$ for complete co-occurrence. Note that hyperparameters, such as the number of topics, will affect the coherence metric. All the experiments for this study are done at a selection for the number of topics that equals five. This number is chosen because it matches the dataset's number of categories.

Topic diversity describes how semantically diverse the obtained topics are. One way to define diversity is introduced in [11]. It defines topic diversity as the percentage of unique words in the top 25 words of all topics. The top 25 words from each topic is appended to a list, which would typically contain repeated words and the percentage of the unique words to the total length of this list is the diversity score. A topic model should generate diverse topics and score high on this metric. A low score indicates redundant topics, which means the model could not sufficiently disentangle the themes in the corpus.

Note that the selection of the number of topics and the vocabulary size in the corpus affects the topic diversity and coherence. This study evaluates the output topics by the NPMI coherence and topic diversity.

3.10 Training Procedure

All models were implemented in python 3.8 using open source libraries OCTIS [19], and BERTopic [12]. Transformers were obtained from the Hugging Face transformers library. All models were trained on a desktop equipped with AMD Ryzen 5 2600 Six-Core Processor with 16 Gigabytes of RAM and a single NVIDIA RTX 2070.

4 Results and Discussion

The obtained model results are summarized in the Table 2. The BERTopic with "symanto/sn-xlm-roberta-base-snli-mnli-anli- xnli" transformer model used to obtain sentence embeddings achieved the highest coherence scores (0.1147) and the second best topics diversity score (0.87).

It is also evident that BERTopic models achieve better performance in general. The reason is that pre-trained transformers provide a dynamic representation of tokens that models polysemous and synonymous tokens better. This representation inherently regularizes the models to generate more coherent topics. The high coherence and diversity is also enhanced by using Maximal Marginal Relevance (MMR) in BERTopic. After obtaining a set of words that describe a group of documents, MMR finds the most coherent words without having too much overlap between the words themselves.

Table 2. Models' results

Model	NPMI	Diversity	Training time (sec)
LDA	0.06275	0.7	15.6
NMF	0.07437	0.86	16.4
CTM	−0.00444	0.92	2685
ETM	0.02571	0.84	3025
BERTopic (AraBERT)	0.08420	0.75	4309
BERTopic (Roberta)	0.11469	0.867	10613

The second BERTopic variant using a sentence transformer to obtain the embeddings outperformed AraBERT. While contextualized word embedding represents a word in a context, sentence encoding represents a whole sentence. Due to the inherent ambiguity in the Arabic language, this higher level of abstraction in looking at the text is advantageous because it allows the models to capture the true meanings better by considering the context. This higher level of abstraction also hides the noisy tokens- those tokens that are not in harmony with the document's topics, which would otherwise misplace a document in a wrong cluster and eventually propagate to topic word representation resulting in reduced coherence.

The time to train the BERTopic models is significantly larger than LDA, NMF, ETM, and CTM. It is worth noting that the time consumed by BERTopic is largely due to obtaining embeddings steps which can be parallelized.

LDA and NMF achieve comparable coherence scores, whereas NMF outperforms LDA in diversity. This agrees with the findings of [16]. Furthermore, they both outperform the more recent neural ETM and CTM models in coherence. Notably, however, CTM achieves high diversity score resulting in disentangled topics. CTM is well known in the literature for generating diverse topics, generally outperforming BERTopic in this regard [12]. We hypothesize that the poor performance of CTM and ETM in coherence is due to the small number of corpus tokens in medium sized documents. This sparsity impacted CTM and ETM but not BERTopic which incorporates higher level of prior knowledge and can handle sparsity better. Although NMF and LDA observe the word co-occurrence patterns to capture the topics, they do this at the document level. In our case,

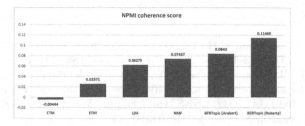

Fig. 3. Coherence scores

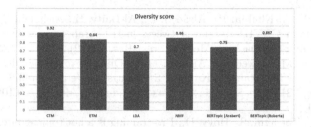

Fig. 4. Diversity scores

the dataset vocabulary size is small but the documents length is not short. This means there is no word sparsity at the document level and hence, their performance was not poor.

It is known in the literature that ETM performance is generally degraded with datasets containing shorter documents [19], which highlights BERTopic superiority as a neural model in dealing with shorter documents and smaller-sized datasets (Figs. 3 and 4).

5 Conclusion

This paper used various topic models to obtain topics of a modern standard Arabic benchmark dataset. LDA and NMF were used to set the scores benchmark. Then three different state-of-the-art neural topic models were used, Contextualized Topic Model (CTM), Embedded Topic Model (ETM), and BERTopic, which uses Bidirectional Encoder Representations from Transformers (BERT). With BERTopic, two transformers were used, Arabert and "symanto/sn-xlm-roberta-base-snli-mnli-anli-xnli": a multilingual siamese network model which achieved the best results. This result is due to the dynamic sentence embeddings that models polysemy and synonymity. The perfromance is contingent on the dataset characteristics, which is preprocessed to include a small number of documents and a small vocabulary size. More work can be done to examine the effect of different datasets exhibiting different characteristics written in different Arabic varieties on the performance of the topic models.

References

1. Abuzayed, A., Al-Khalifa, H.: BERT for Arabic topic modeling: an experimental study on BERTopic technique. Proc. Comput. Sci. **189**, 191–194 (2021)
2. Al Qudah, I., Hashem, I., Soufyane, A., Chen, W., Merabtene, T.: Applying latent Dirichlet allocation technique to classify topics on sustainability using Arabic text. In: Arai, K. (ed.) SAI 2022. LNNS, vol. 506, pp. 630–638. Springer, Cham (2022). https://doi.org/10.1007/978-3-031-10461-9_43
3. Alhaj, F., Al-Haj, A., Sharieh, A., Jabri, R.: Improving Arabic cognitive distortion classification in Twitter using BERTopic. Int. J. Adv. Comput. Sci. Appl. **13**(1), 854–860 (2022)
4. Alshalan, R., Al-Khalifa, H., Alsaeed, D., Al-Baity, H., Alshalan, S.: Detection of hate speech in COVID-19-related tweets in the Arab region: deep learning and topic modeling approach. J. Med. Internet Res. **22**(12), e22609 (2020)
5. Alshammeri, M., Atwell, E., Alsalka, M.A.: Quranic topic modelling using paragraph vectors. In: Arai, K., Kapoor, S., Bhatia, R. (eds.) IntelliSys 2020. AISC, vol. 1251, pp. 218–230. Springer, Cham (2021). https://doi.org/10.1007/978-3-030-55187-2_19
6. Bianchi, F., Terragni, S., Hovy, D.: Pre-training is a hot topic: contextualized document embeddings improve topic coherence. In: Proceedings of the 59th Annual Meeting of the Association for Computational Linguistics and the 11th International Joint Conference on Natural Language Processing (Volume 2: Short Papers), pp. 759–766. Association for Computational Linguistics (2021)
7. Bianchi, F., Terragni, S., Hovy, D., Nozza, D., Fersini, E.: Cross-lingual contextualized topic models with zero-shot learning. In: Proceedings of the 16th Conference of the European Chapter of the Association for Computational Linguistics: Main Volume, pp. 1676–1683. Association for Computational Linguistics (2021)
8. Blei, D.M., Ng, A.Y., Jordan, M.I.: Latent Dirichlet allocation. J. Mach. Learn. Res. **3**, 993–1022 (2003)
9. Cao, Z., Li, S., Liu, Y., Li, W., Ji, H.: A novel neural topic model and its supervised extension. In: Proceedings of the AAAI Conference on Artificial Intelligence, vol. 29, no. 1 (2015)
10. Deerwester, S., Dumais, S.T., Furnas, G.W., Landauer, T.K., Harshman, R.: Indexing by latent semantic analysis. J. Am. Soc. Inf. Sci. **41**(6), 391–407 (1990)
11. Dieng, A.B., Ruiz, F.J.R., Blei, D.M.: Topic modeling in embedding spaces. Trans. Assoc. Comput. Linguist. **8**, 439–453 (2020)
12. Grootendorst, M.: BERTopic: neural topic modeling with a class-based TF-IDF procedure. Technical report arXiv:2203.05794, arXiv (2022)
13. Miao, Y., Grefenstette, E., Blunsom, P.: Discovering discrete latent topics with neural variational inference. In: ICML (2017)
14. Newman, D., Lau, J.H., Grieser, K., Baldwin, T.: Automatic evaluation of topic coherence. In: Human Language Technologies: The 2010 Annual Conference of the North American Chapter of the Association for Computational Linguistics, pp. 100–108. Association for Computational Linguistics, Los Angeles (2010)
15. Obeid, O., et al.: CAMeL tools: an open source python toolkit for arabic natural language processing. In: Proceedings of the 12th Language Resources and Evaluation Conference, pp. 7022–7032. European Language Resources Association, Marseille (2020)
16. O'Callaghan, D., Greene, D., Carthy, J., Cunningham, P.: An analysis of the coherence of descriptors in topic modeling. Expert Syst. Appl. **42**(13), 5645–5657 (2015)

17. Rafea, A., GabAllah, N.A.: Topic detection approaches in identifying topics and events from Arabic corpora. Proc. Comput. Sci. **142**, 270–277 (2018)
18. Schofield, A., Magnusson, M., Mimno, D.: Pulling out the stops: rethinking stopword removal for topic models. In: Proceedings of the 15th Conference of the European Chapter of the Association for Computational Linguistics: Volume 2, Short Papers, pp. 432–436. Association for Computational Linguistics, Valencia (2017)
19. Terragni, S., Fersini, E., Galuzzi, B.G., Tropeano, P., Candelieri, A.: OCTIS: comparing and optimizing topic models is simple! In: Proceedings of the 16th Conference of the European Chapter of the Association for Computational Linguistics: System Demonstrations, pp. 263–270. Association for Computational Linguistics (2021)

Modelling

Architectural Invariants and Correctness of IoT-Based Systems

Christian Attiogbé[(⊠)] and Jérôme Rocheteau

Nantes Université, École Centrale Nantes, CNRS, LS2N, UMR 6004, 44000 Nantes, France
{christian.attiogbe,jerome.rocheteau}@ls2n.fr

Abstract. Systems based on the Internet of Things impact more and more industrial areas such as smart manufacturing, smart health monitoring and home automation. Ensuring their correct construction, their well functioning and their reliability is an important issue for some of these systems which can be critical in case of dysfunction. The main requirements on physical architectures and control software are common in most of IoT-based systems. Therefore, we propose on the basis of their common architectural properties and behaviour, a generic formal model of IoT-based systems together with the rigorous analysis of their consistency properties; specific properties may be gradually added and checked. The proposed generic formal model is implemented as a parametrised model and experimented using the Event-B framework. This parametrised model is extensible; it can be profitably adapted to more general hybrid or cyber-physical systems. Moreover, our generic model is independent of the target formal modelling tools, it can be implemented in various other formal analysis environments.

Keywords: IoT Applications · Generic formal model · Invariant properties · Event-B

1 Introduction

Internet of Things *systems* impact industrial areas such as smart manufacturing, smart vehicles, smart logistics and transportation, smart farming, etc. These applications share a well-established architectural structuring, reference models and some functional and non-functional properties [2,6,8]. However, well-established engineering methods and techniques are still needed [19] to ensure that the applications are reliable, secure, scalable, well-integrated, and extensible. In this context, we are motivated by proposing methods and tools for mastering the modelling, the analysis and the development of such IoT-based applications. The challenge is that these applications can become rapidly complex because of their heterogeneous and evolving environment.

In order to ensure the consistency and the well-functioning of an IoT-based application, the latter should integrate as a parameter, the complete description of the physical context that it controls. Therefore, a global model can be built and analysed with respect to intrinsic consistency and to the required specific properties. We propose such a global formal modelling and the related analysis.

P. Fournier-Viger et al. (Eds.): MEDI 2022, CCIS 1751, pp. 75–88, 2022.
https://doi.org/10.1007/978-3-031-23119-3_6

The contributions of this paper are manifold: we propose *i)* a generic formal description of the physical architecture of an IoT-based system; *ii)* a formal description of the control application parametrised by its physical environment; *iii)* the modelling and analysis of the invariant architectural properties of such IoT-based systems and the description of some specific properties. They are described to be customizable for other application cases. We design using Event-B, a formal parametrised model, to support the full modelling and analysis approach. Moreover, we build a tool that generates systematically for a given IoT-based application, the specific Event-B parts that are used to instantiate the parametrised Event-B model.

The organisation of the article is as follows. In Sect. 2, we introduce the background concepts. Section 3 is devoted to the generic formal model of IoT-based applications. In Sect. 4, we deal with the invariant consistency properties, formalised with operational semantic rules. In Sect. 5, we show how we have implemented our proposed generic formal model and analysis technique using Event-B. We compare our work to the related ones in Sect. 6. Finally, Sect. 7 presents some perspectives and future work.

2   Basic Concepts and Architecture of IoT Systems

Based on state-of-the-art references [2, 8, 14, 20] of IoT technologies, challenges, comparisons, and reference models, we consider the following main elements of IoT.

A *thing*[1] is a physical object equipped with: *i)* sensors that collect and gather data from the environment; *ii)* actuators that allow the control of the *thing* or allow the *thing* to act on its environment.

An IoT-based *application* is a software made of services, built on top of the physical infrastructure made of one or several things. An example of the architecture of a control application is depicted in Fig. 1 where we can distinguish: *i)* a physical part made of the controlled devices equipped with sensors and actuators; *ii)* a software part made of the (sub-)controllers which interact with the physical part through an event dispatcher. This abstraction covers the four-layers architecture widely admitted [2] now for service-oriented architecture (SOA) IoT systems.

A *control application* sends orders (including signals) to actuators, according to information collected by sensors. A control application often uses *rules* stated by a human expert or systematically computed from a specific database, to issue control orders. Therefore, the main components to deal with for an IoT-based system are: a physical part made of sensors, actuators, *things*, and network infrastructure; a software part made of a *control application* and potentially specific control or monitoring services. Additionally, the components interact through low level or application level communication protocols such as WiFi, bluetooth, ZigBee and MQTT [4].

3   Formal Model of IoT-Based Systems

We propose in the following an abstract formal model of IoT-based systems. We use set theory and relation notations to structure the model components. Sets are written with

---

[1] we keep the vocabulary of IoT domain.

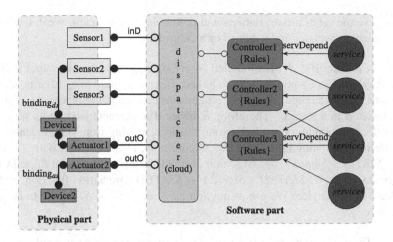

Fig. 1. An abstract architectural view of an IoT-based system

capital letters, the standard set operators ($\in, \subseteq, \cdots$) are used. A relation r defined over the sets S and P is written: $r : S \leftrightarrow P$ or more conventionally $r \subseteq S \times P$; a function f over S and Q is written: $f : S \rightarrow Q$; the relational operators ran and dom denote respectively the *range* and the *domain* of a relation or a function.

An IoT-based application is composed at least of: a set of connected IoT devices ($\mathcal{D}$), sensors ($\mathcal{S}$) and actuators ($\mathcal{A}$), that make the physical part; a set of controllers (C), sometimes together with a dedicated server (or dispatcher) which collects the data from sensors and distributes them to the controllers. Sensors and actuators are bound to the devices. A controller is linked to the sensors from which it reads inputs, and to the actuators it manages. Sensors and actuators may be connected to several interconnected controllers.

3.1 Elementary Components of the Model

Sensors. A sensor s provides a value in a given range; a value will correspond to a state of the device that it senses. Each category (c_s) of sensors may have various value ranges (R_{cs}). A sensor interacts with its environment through communication protocols. Let *CommProto* be such a set of protocols; a sensor s of the set of sensors $\mathcal{S}$ ($s \in \mathcal{S}$) is defined by a 4-tuple ($categ_s, range_{sc}, value_s, comm_{p_s}$) with a category $categ_s \in C_S$, a range $range_{sc} \subseteq R_{cs}$, a current value in its range $value_s \in r_{sc}$ and a set of communication protocols $comm_{p_s}$. We use the following relations to get each element of the 4-tuple:

$$categ_s : S \rightarrow C_S \quad range_s : S \leftrightarrow R_{cs}$$
$$value_s : S \rightarrow R_{cs} \quad comm_{p_s} : S \leftrightarrow CommProto$$

Actuators. An actuator a of the set of actuators $\mathcal{A}$ ($a \in \mathcal{A}$) receives an input in a specific range of order values ($Order_A$), and delivers accordingly an output signal (from a set $Signal_A$) towards its environment. We use the triple ($inputOrd_a, ouputSign_a, comm_{p_a}$) to describe an actuator and the following relations to determine its elements:

$$inputOrd_a : \mathcal{A} \leftrightarrow Order_A \quad ouputSign_a : \mathcal{A} \leftrightarrow Signal_A \quad comm_{p_a} : \mathcal{A} \leftrightarrow CommProto$$

They give the set of inputs, outputs and protocols of an actuator. We define later the links between an actuator and a device or a controller.

Devices. A device $(d \in \mathcal{D})$ is modelled by its state s_d in such a way that a range of measured values V_v of some sensors s (linked d), corresponds to s_d (that is, $V_v \mapsto s_d$). In the same way, the output signals of an actuator can, upon the reception of an order, set the device d in a state s_d. Therefore, a device d is characterised by its set of states (*STATED*), corresponding to its behaviour which is, moving from state to state according to the received stimulation signals. Accordingly, d is modelled with a transition system $\langle STATED, SignalA, S_0, \delta \rangle$ which abstracts its behaviour; with *SignalA* the set of received signals, $\delta : STATED \times SignalA \rightarrow STATED$ a transition function, and S_0 an initial state of the device. We use a function $curState_d : \mathcal{D} \rightarrow STATED$ to denote the current state.

Services. The tasks performed in controller applications consist in applying a set of control rules (R) to analyse data (D_S) collected from the sensors (S) and to compute, the orders (*Order*) to be sent to the actuators. The services implement these control rules. A controller c is then equipped with a function $ComputeOrder_R : D_S \rightarrow Order$. When the controller collects values from sensors bound to it, it computes the appropriate order using the rules R, and outputs this order to the actuator bound to it.

3.2   Abstractions for IoT-Based Systems

We describe an abstract model of an IoT-based system according to the two main components presented in Sect. 2: we build the abstract model M_{phys} of the physical part and the abstract model M_{soft} of the control software part. We consider the previous sets $\mathcal{S}$, $\mathcal{A}$ and $\mathcal{D}$. The physical architecture is modelled with a n-tuple $M_{phys} = (S, A, D, binding_{(d,s)}, binding_{(a,d)})$ where $S \subseteq \mathcal{S}$ is a subset of sensors; $A \subseteq \mathcal{A}$ is a subset of actuators; $D \subseteq \mathcal{D}$ is a set of sensed or controlled devices; $binding_{(d,s)} \subseteq D \times S$ is a relation that describes the binding between the sensed devices and their sensors, and $binding_{(a,d)} \subseteq A \times D$ is a relation that describes the binding between the actuators and the controlled devices. The control dependency between devices is defined later.

A control software (or application) part of an IoT system is modelled with the tuple $M_{soft} = (C_R, Serv, servDepend_{(c,s)})$ where $C_R \subseteq C$ is a set of controllers which use a set of the control rules R for their control tasks; *Serv* is a set of services used or provided by the controllers; $servDepend_{(c,s)} \subseteq C_R \times Serv$ is the dependencies between the controllers and the used services.

To model the link between the control application and the sensors and actuators, the physical (M_{phys}) and software (M_{soft}) models are linked with the following relations:

- $inD \subseteq S \times C_R$ which models the links between the sensors and the controller; it supports data input from the sensors;
- $outO \subseteq C_R \times A$ which models the links between the controller and the actuators; it supports order output to actuators.

At this stage the control application (M_{soft}) can communicate with the physical part via its abstract model (M_{phys}). Typically the application receives data from the sensors (via *inD*) and issues orders sent to the actuators (via *outO*). The orders are computed from the application services using defined rules.

But, in order to state the *architectural invariants* and to analyse the IoT system properly, we need to address one of the identified defects leading to inconsistencies, which is the lack of explicit declaration of dependencies between sensors and related controlled devices. When the control of a given device depends on some sensors, this dependency should be made explicit. The devices should have been equipped by an actuator which share the same controller with the involved sensors. Therefore, it is necessary to make explicit in the model, the control dependency relation between involved sensors and controlled devices; this is done with a relation $CtrlDepend_{(s,d)} \subseteq S \times D$.

This relation describes which sensors are used to control which devices, so that we can reason later on the consistency of the functioning of the global system. Both the relations $CtrlDepend_{(s,d)}$ and $binding_{(d,s)}$ are necessary since the impacted devices described by $CtrlDepend_{(s,d)}$ can be different from the sensed ones described by $binding_{(d,s)}$. Consequently, given a control part $M_{soft} \cong (C_R, Serv, servDepend_{(c,s)})$ with a physical architecture $M_{phys} \cong (S, A, D, binding_{(d,s)}, binding_{(a,d)})$, and their interconnection with the relations *inD*, *outO* and $CtrlDepend_{(s,d)}$, the model of the complete control system *Sys* integrating them is described by the 5-tuple:

$$Sys = (M_{phys}, M_{soft}, inD, outO, CtrlDepend_{(s,d)})$$

For the systematic construction and analysis of the model, we structure it with parameters made of all, or some parts, of the global system; hence the genericity; different parameters lead to specific systems: $Sys[M_{phys}, M_{soft}, inD, outO, CtrlDepend_{(s,d)}]$. This enables us to build separately the different parts, and also to modify them easily as well as their interconnections; we can freeze a physical architecture and check different versions of the control part or as done in the following, freeze the software and check some configurations of the physical parts. Then, the global model is a parametrised structure with the parameters M_{phys}, *inD*, *outO* and $CtrlDepend_{(s,d)}$, and a frozen software part:

$$Sys_{(M_{soft})}[M_{phys}, inD, outO, CtrlDepend_{(s,d)}]$$

3.3 Behavioural Description of a Control Application

A control application continuously reacts to the data collected by sensors and changes the state of its environment by sending orders to the involved actuators. We use operational semantic rules to describe this general behaviour. First, we assume that the application is *consistent* so that it can react properly to the sensed data. In Sect. 4.1 we show how the consistency properties are defined and then how they can be checked in Sect. 4.2. Given $Sys_{(M_{soft})}[M_{phys}, inD, outO, CtrlDepend_{(s,d)}]$, a consistent application with $M_{phys} = (S, A, D, binding_{(d,s)}, binding_{(a,d)})$ and $M_{soft} = (C_R, Serv, servDepend_{(c,s)})$, when a controller c_i (with a function $computeOrder_{c_i}$) of C_R, receives a value val from a sensor s_i of S bound to a device d_s of D, considering that the control of a device d_c depends on the sensor s_i, and that there is an actuator a_s bound to d_c, then an order ord_i, computed by the controller c_i linked to a_s, is sent to the actuator a_s.

Control Abstraction and Genericity. A sensor covers a set of ranges that will be matched to a set of states of the device. We consider this matching, as a (explicitly provided) function σ from a set of values to a set of states. Consequently, given a sensor s_i with its range of values r_{sc}, a device d_c characterised by $\langle STATED, Signal, S_0, \delta \rangle$, in order to control d_c, we need a function $\sigma_{sd} : r_{sc} \rightarrow STATED$ (considered as a *parameter* of a controler) that links s_i and d_c. This is provided through the services on which the controller depends. Therefore, a controller c_i is parametrised by s_i, d_c and σ_{sd} (denoted by $c_i[s_i, d_c, \sigma_{sd}]$) in such a way that given a value v from s_i, $\sigma_{sd}(v)$ provides the corresponding state s of the device d_c. Before changing the state of d_c according to values sensed by s_i and the predefined behaviour δ of d_c, the controller should compute, provided that σ_{sd} is already defined through its services, the appropriate order to be sent to the device. Hence, the following generic semantic rule (compOrd) to compute orders. Note that, when $curState(d_c) = s$, nothing should be changed.

$$\frac{\begin{array}{c} s_i \mathrel{\widehat{=}} (c_s, r_{sc}, v_s, comm_{p_s}) \\ d_c \mathrel{\widehat{=}} \langle STATED, Signal, S_0, \delta \rangle \\ \sigma_{sd} : r_{sc} \rightarrow STATED \qquad v_s \in r_{sc} \\ \sigma_{sd}(v_s) = s \qquad curState(d_c) \neq s \\ \mathsf{thisOrder} \in Signal \ \ (curState(d_c), \mathsf{thisOrder}, s) \in \delta \end{array}}{ComputeOrder_{c_i[s_i, d_c, \sigma_{sd}]}(v_s) = \mathsf{thisOrder}}\text{(compOrd)}$$

We formally define the general behaviour of the IoT application by the following operational semantic rule where the operators $\Downarrow$ and $\Uparrow$ denote respectively the reception of a value from a given sensor by a controller and the sending of an order by a controller to an actuator. Thus $c_i \Downarrow (s_i, \mathsf{val})$ expresses that the controller c_i receives the value val sent by the sensor s_i; similarly $c_i \Uparrow (a_s, \mathsf{ord}_i)$ expresses that the controller c_i sends the order ord_i to the actuator a_s. The rule (RoS - react on sense) captures the traditional sense-decision-control paradigm of control systems.

$$\frac{\begin{array}{c} M_{soft} \mathrel{\widehat{=}} (C_R, \ Serv, \ servDepend_{(c,s)}) \\ M_{phys} \mathrel{\widehat{=}} (S, A, D, binding_{(d,s)}, binding_{(a,d)}) \\ Sys_{(M_{soft})}[M_{phys}, \ inD, \ outO, \ CtrlDepend_{(s,d)}] \\ s_i \in S \quad a_s \in A \quad d_s \in D \quad d_c \in D \quad c_i[s_i, d_c, \sigma_{sd}] \in C_R \\ binding_{(d,s)}(d_s) = \{s_i\} \qquad (a_s, d_c) \in binding_{(a,d)} \\ (s_i, c_i) \in inD \quad (c_i, a_s) \in outO \quad (s_i, d_c) \in CtrlDepend_{(s,d)} \\ c_i \Downarrow (s_i, \mathsf{val}) \quad \mathsf{val} \in range_s(s_i) \\ \mathsf{ord}_i = ComputeOrder_{c_i[s_i, d_c, \sigma_{sd}]}(\mathsf{val}) \end{array}}{c_i \Uparrow (a_i, \mathsf{ord}_i)}\text{(RoS)}$$

A consequence of the RoS rule is the *Integrity of orders*: any order sent to an actuator results from one of the services (they provide σ_{sd}) of the control application. Since the computation of orders is due to the controller whose services implement (via σ_{sd}) the control application rules (previously denoted R), the orders sent to the actuators are the right ones. However, the integrity checking should be propagated until the application implementation level.

The RoS rule expresses one step of the cyclic behaviour of the control application; the repetition of the step is captured by the continuous enabling of the rule. We have also generalized the RoS rule by considering an extension of $computeOrder_{c_i}$, with a function $ComputeGlobalOrder_{Sys}$ that computes an order, according to the global state of the system. A design flaw may happen at the system level; we formalise this situation with a semantic rule (dFlaw rule) which expresses that when there is no transition from the current state of the controlled device d_c to a state corresponding to the value sent by a related sensor (i.e. $(curState(d_c), *, s) \notin \delta$), the controller is not able to send an order to the actuator, since the order is undefined. In the case of inconsistent values we have to implement (based on the dFlaw rule) a rule that raises inconsistency of sensors.

Further Generalisation. The rule compOrd considers the computation of an order related to the state of one device; this rule may be extended to the states of several devices, and also the computation of several orders to be emitted to change the state of the system; it will involve an evolution with a sequence of orders. This requires the extension of the emission operator to the emission of the sequence of orders.

From now on, we have captured the correct behaviour and the possible design flaw of an IoT-based control application.

4   Consistency Properties and Analysis of the Formal Model

The generic formal model built in the Sect. 3 is enhanced with consistency properties.

4.1   Invariant Consistency Properties

We focus on the architectural consistency and then, on the consistency of the functioning of IoT-based applications. Let us consider in the following, an application defined with

$$M_{phys} \stackrel{\frown}{=} (S, A, D, binding_{(d,s)}, binding_{(a,d)}) \qquad M_{soft} \stackrel{\frown}{=} (C_R, Serv, servDepend_{(c,s)})$$
$$Sys_{(M_{soft})}[M_{phys}, inD, outO, CtrlDepend_{(s,d)}]$$

We describe a list of properties required for an IoT architecture to be consistent so that a model satisfying these properties will be consistent.

Well-Structuring of Physical Components (wsHW). An IoT architecture involves devices, sensors and actuators; it is described with two binding relations. To be controlled, the architecture requires the following property which expresses partial connectivity (with a disjunction) in order to be less restrictive, instead of full connectivity: $binding_{(d,s)} \neq \varnothing \ \vee \ binding_{(a,d)} \neq \varnothing$

Well-Structuring of Controllers (wsCtrl). An IoT control application requires a connection with at least one sensor and one actuator: $\forall c_i \in C_R.(inD(c_i) \neq \varnothing \ \wedge \ outO(c_i) \neq \varnothing)$

Weak Consistency of Components Involved in Interactions (wkCst). For consistency purpose, sensors or actuators involved in the interactions should be those described in the physical support: $dom(inD) \subseteq S \ \wedge \ ran(outO) \subseteq A$

However, this consistency is weak, because it does not constrain the linking of the involved sensors and actuators. For more accuracy, we define the following stronger property.

Consistency of Control Dependencies (CstBind). The consistency of the device controls requires that: the actuators to whom a controller sends its orders (via *outO*), are those actuators bound (via $binding_{(a,d)}$) to the devices which are controlled (via $CtrlDepend_{(s,d)}$) by the sensors bound (via *inD*) to the controller. Thus and interaction consistency property is required; it is expressed by the equality of the composition of the involved relations: $inD; outO = ctrDepend; binding_{ad}^{-1}$

The *Consistency of control dependencies* property ensures that: if a sensor s_i impacts the control of a given device d_c (via $CtrlDepend_{(s,d)}$), and the sensor s_i is connected to a controller c_i (via *inD*) then the actuator a_k bound (via $binding_{(a,d)}$) to the device d_c is also linked (via *outO*) to the controller c_i.

$$\frac{\begin{array}{cccc} s_i \in S & c_i \in C_R & a_k \in A & d_c \in D \\ (s_i, c_i) \in inD & (a_k, d_c) \in binding_{(a,d)} \\ (s_i, d_c) \in CtrlDepend_{(s,d)} \end{array}}{(c_i, a_k) \in outO} \text{(CstDep)}$$

Well-Structured Connection of Actuators and Sensors (wsS2A). The controllers which are connected to sensors should also be connected to some actuators, otherwise the collected inputs are not used for the control: $ran(inD) \subseteq dom(outO)$

This property can be relaxed if one considers applications without actuators, or with a pool of interacting controllers.

Consistency for Communication Protocols (wsProt). The consistency of protocols requires that the pairs of sensor-controller and controller-actuator use compatible communication protocols: each sensor interacts with the bound controller using an appropriate communication protocol; each controller interacts with the bound actuators using an appropriate communication protocol. Consider the set of communication protocols (*CommProto*) used by the components of the architecture and $c_i \in C_R$, $s_i \in S$, $a_s \in A$, $(s_i, c_i) \in inD$, $(c_i, a_s) \in outO$ such that $comm_{p_s}(s_i) \subseteq CommProto \wedge comm_{p_c}(c_i) \subseteq CommProto \wedge comm_{p_a}(a_s) \subseteq CommProto$. The protocol consistency requires:

$$(comm_{p_s}(s_i) \cap comm_{p_c}(c_i) \neq \varnothing) \wedge (comm_{p_c}(c_i) \cap comm_{p_a}(a_i) \neq \varnothing)$$

4.2   Consistency Analysis of IoT-Based Control Applications

Given the previous consistency properties, we state the following two definitions for the consistency analysis of IoT-based control applications. If a given model satisfies the properties then it is consistent.

Definition 1. *(Architectural correctness) A given physical architecture M_{phys} is said consistent if it preserves the property* wsHW.

The physical architecture can be given without any controller, while the other properties involve the relations with the controller and a specific connection of the architectures.

Definition 2. *(Behavioural correctness) An application* $Sys_{(M_{soft})}$, *parametrised with* M_{phys}, *inD, outO and* $CtrlDepend_{(s,d)}$ *is consistent if:* M_{phys} *is consistent, and if the properties* wsCtrl, wsS2A, wsProt, wkCst, CstBind, CstDep *hold.*

A formal model of an application, which has these properties established, will be consistent by construction. This is the main idea implemented in the following section.

5 Checking the Consistency Properties Using Event-B

The abstract and generic model and properties defined in Sects. 3 and 4 should be implemented and analysed in a given formalism with its related tools, for any specific IoT-based application. We propose, using the Event-B formalism, a *parametrised Event-B model* as a generic base to model and analyse IoT-based applications. This parametrised approach is more interesting and more general than the straight translation or implementation of the abstract formal model in Event-B or in any other formal language; indeed, only a few part of the abstract model is specific to a given application, the remaining major part (is common to all applications and should) stay unchanged.

Overview of Event-B Modelling. Event-B [1,10] is a modelling and development method where components are modelled as *abstract machines* which are composed and refined into concrete machines called *refinements*. An *abstract machine* comprises a state space invariant and guarded events; it describes a mathematical model of a system behaviour as a discrete transition with the guarded events. *Proof obligations* are defined to establish model consistency via invariant preservation. Specific properties (included in the invariant) of a system are also proved in the same way.

5.1 A Parametrised Model for Consistency Checking of IoT Applications

We capture the common requirements and properties of IoT-based applications within an abstract and generic model; the analysis of the consistency properties does not depend on a specific application; it may be done through a generic parametrised model. We implement the generic model[2] in Event-B following the structure of a *parametrised model* interconnecting in a systematic way, a physical part and a control software part.

This justifies the structuring of our Event-B model where some *contexts* and *machines* are to be adapted to specific applications but other machines/contexts are defined once for all. The parametrised model, as a composition of Event-B components (see Fig. 2), is not only designed and used to implement our proposed method of modelling and analysis, it aims at being an easily reusable base. For this purpose, we adopted a layered structuring of the Event-B components in order to have a systematic approach for building, generating or extending the generic model. For extensibility, we consider categories of IOT-based systems; for instance, a home control category where the main components of systems are always the same: lights, windows, doors, garage,

---

[2] The complete Event-B development can be found at https://gitlab.univ-nantes.fr/attiogbe-c/iot_with_eventb.

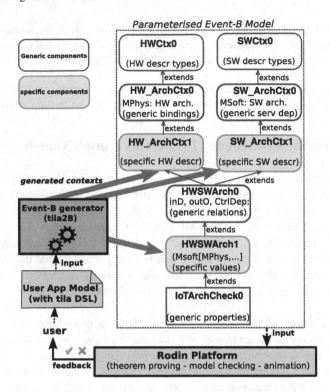

Fig. 2. The architecture of our generic Event-B modelling and analysis framework

heating, etc. Therefore the modelling components are not varying and can be gathered as reusable components in our formal modelling.

The generic architecture is depicted in Fig. 2. A first basic layer comprises fixed predefined Event-B contexts (whose names end with 0) which gather all elementary types and relations required in any application within a given category. Another layer comprises Event-B contexts and machines (whose names end with 1) which are the specific instantiations of the predefined contexts. Considering the category of home automation applications, at the hardware level, the context HWCtx0 contains the basic sets (LIGHTSENSOR, MOTIONSENSOR, LIGHTACTUATOR, etc.); at the software level the context SWCtx0 contains all the basic sets (SERVICE, CONTROLLER) for the applications of this category.

The context HW_ArchiCtx0 implements M_{phys}; it contains the generic structuring of a physical architecture (the formal bindings between the devices); that is, the relations $binding_{(d,s)}$ and $binding_{(a,d)}$ (see Sect. 3.2); similarly SW_ArchiCtx0 implements M_{soft}; it contains the generic structuring of the software part (with *servDepend*).

The context HW_ArchiCtx1 contains a specific instantiation for the physical architecture; it comprises the declarations of the objects (the sensors of each type, the needed controllers, etc.) and their assembly in a given application. Similarly, SW_ArchiCtx1 contains a specific instantiation of the software part; it comprises the controllers and the services on which they depend. Thus, only these two contexts will be modified to consider new instances of physical or software part. The generic interconnection between the two parts (with the relations *inD*, *outO*, *CtrlDepend*, see Sect. 3.2) is implemented with the Event-B context HWSW_Archi0. In the same way, the context HWSW_Archi1 models a specific instantiation; it is the only context to be modified in order to build an interconnection for a specific control application; it gathers M_{soft} and the four parameters M_{phys}, *inD*, *outO*, *CtrlDepend*.

The analysis machine (IoTArchiCheck0, detailed in Fig. 3) is defined once for all; it contains the invariant properties (wsHW, wkCst, CstBind, CstDep, wsS2A defined in Sect. 4.1) to be checked for any given IoT-based application.

Each specific application is given as a parameter (HWSW_Archi1) of IoTArchiCheck0; that justifies the structuring with the Event-B SEES clause: a way to implement the genericity. Note that the effective implementation of an example system, is orthogonal to the preliminary step of consistency checking. If the IoTArchiCheck0 machine parametrised with HWSW_Archi1 is proved correct, then all the architectural and consistency properties are satisfied and consequently the related model is consistent.

```
MACHINE IOTArchiCheck0
SEES HWSWArchi1
INVARIANTS
  wsHW: ⟨theorem⟩ (binding_ds ≠ {}) ∧ (binding_ad ≠ {})
  wsS2A: ran(inD) ⊆ dom(outO)
  wkCtrl: dom(inD) ⊆ SENSOR ∧ ran(outO) ⊆ ACTUATOR
  CstBind: (inD; outO) = (ctrlDepend; (binding_ad⁻¹))
  wsCtrl: (inD ≠ ∅ ∧ (dom(SENSOR ◁ inD) ⊆ SENSOR)) ∨
          (outO ≠ ∅ ∧ (dom(CONTROLLER ◁ outO) ⊆
          CONTROLLER))
EVENTS                      ...
END
```

Fig. 3. The generic analysis Event-B machine

The aim is now, given a model describing an IoT-based application, to prove at least all the properties of interest listed above. We use Rodin to discharge these proofs. But, the users can add other specific properties as needed. To facilitate the use of our method, it is supported by a tool-assisted process (where the contexts HW_ArchiCtx1, SW_ArchiCtx1, HWSW_Archi1 are generated) that is presented below.

5.2 Putting into Practice and Assessment

To ease the modelling and the analysis of IoT-based applications using our approach, we design a process to support it and to facilitate its reuse. We define τila, a tiny IoT domain specific language that helps to describe any application, by defining the used objects, their relations and their behaviours. Then we build a tool τila2B[3], that uses as

---

[3] https://gitlab.univ-nantes.fr/attiogbe-c/iot_with_eventb.

input the description in the τ*ila* DSL and generates the Event-B models as the specific parameters (HW_ArchiCtx1, SW_ArchiCtx1, HWSW_Archi1) to be plugged in our parametrised model, (see Fig. 2); the resulting Event-B model can then be submitted to Rodin for analysis. We demonstrate our generic model and the process with several examples. As for the analysis process, if the machine IoTArchiCheck0 is proved correct using Rodin then the model used as parameter is consistent.

6   Related Work

Existing related often take into account a specific concern (QoS, security, time performanxe, protocols, etc.), and as such they can be considered as complementary. But, in our knowledge there is no widely shared abstract model that can help the interoperability between the existing proposals and results. We target this objective by proposing, compared to some of the existing works, an open and extensible abstract model. In [20] the authors compare state-based and rule-based models of smart home apps for analysis scalability purpose. Their models are not generic and they focus on detecting misleading coordinations of components. A comprehensive survey in [9], emphasizes security aspects, proposes a hybrid security analysis system, but also shows that few attention are paid to abstract models and verification aspects. In [5] the authors introduce SysML4IoT to define a model compliant with the IOT-A reference model, and they translate the SysML model into NuSMV programs for the analysis concern. Their focus was on the verification of quality of service (QoS) properties. The authors of [14] focus on a multiview modelling together with workflows for implementing cloud-based Industrial IoT systems. For the modelling they combine several views through various models, and integrate them using the Automation Markup Language; they chose Uppaal for verification aspects and combine the Uppaal Timed Automata models with action patterns of timing behaviour to verify the timing performance to guarantee timing properties. The concerns of [7, 16] are related to IoT services for health-care. In [15] the authors propose a development methodology and an associated framework to ease the development of IoT applications, but formal analysis was not their concern. In [3, 11] verification of communication protocols such as MQTT are dealt with; Timed process-algebra [3] and Probabilistic timed automata and statistical model checking [11] are used for this purpose. In [13] the authors focus on the verification of the correct composition of IoT objects described as labelled transitions equipped with input and output interfaces. The objects composition results in a synchronous composition of the objects LTS leading to a composite service; then the notion of (concurrent processes) compatibility is used to ensure that the composite service has a correct interaction of its component services. Well-composed objects are then deployed on the basis of their mutual dependency. We share the formal modelling and verification objectives with [13], but our approach is more focused on building correct control-oriented IoT-applications; while their composition is restricted to binary parallel composition of object behaviours, we propose a more flexible global interaction based on the sense-decision-control paradigm which ensures flexible n-ary composition. We do not deal with deployment aspects, we rather provide means to generate the model, to analyse and simulate the targeted applications.

7 Conclusion

We have designed an abstract formal model of IoT-based applications learning from the common properties of IoT-based systems. An IoT-based application is then structured in a generic way by distinguishing different parts related to the physical part, the software part and the interconnection relations between both parts. These different parts serve as parameters in order to favour extension and reusability of the abstract formal model. We make explicit in the model, the architectural and consistency properties that have been formalised as the invariants of many IoT-based applications. These properties are systematically checked during the construction of an application. We then proposed a parametrised Event-B model as an implementation of our generic abstract model. The Event-B implementation is used for experiments that confirm the effectiveness of the proposed approach. To facilitate the modelling of IoT-based application and the reuse of our analysis approach, we design a tiny IoT application description language (τila) and a tool that generates for a given application expressed in τila, the Event-B components to be used to instantiate the parametrised model; thus the process is fully automatised.

Perspectives. Considering that IoT-based applications are a subset of cyber-physical systems, mostly characterised by their heterogeneous features, the proposed method here may be generalised to heterogeneous systems. For this purpose, we plan to connect our framework with existing DSLs which enable one to describe IoT systems; their descriptions will thus benefit from the formal modelling and the rigorous analysis of the designed systems prior to implementation. We already identified such DSLs for further investigation: openIoT [12], SDL-IoT [18], ide4dsl [17], UML4IoT [21], SysML4IoT [5, 14], OpenHAB[4].

In order to ensure the robustness of the physical part, both its abstract model and its physical implementation may coexist during the live of the IoT system; both interacting with the sensors and actuators environment, in order to anticipate defects and also security issues. The abstract model, extended for these needs, will then behave as the digital twin of the real system and will enable one to check and to monitor it. Such interaction between models of different abstraction levels is planned for future work.

References

1. Abrial, J.-R.: Modeling in Event-B: System and Software Engineering. Cambridge University Press, Cambridge (2010)
2. Al-Fuqaha, A.I., Guizani, M., Mohammadi, M., Aledhari, M., Ayyash, M.: Internet of things: a survey on enabling technologies, protocols, and applications. IEEE Commun. Surv. Tutorials 17(4), 2347–2376 (2015)
3. Aziz, B.: A formal model and analysis of the MQ telemetry transport protocol. In 2014 9th International Conference on Availability, Reliability and Security, pp. 59–68 (2014)
4. Banks. A., Gupta, R.: MQTT Version 3.1.1Plus Errata 01. http://docs.oasis-open.org/mqtt/mqtt/v3.1.1/mqtt-v3.1.1.pdf. OASIS Standard Inc. (2015)
5. Costa, B., Pires, P.F., Delicato, F.C., Li, W., Zomaya, A.Y.: Design and analysis of iot applications: a model-driven approach. In 2016 IEEE 14th International Conference on Dependable, Autonomic and Secure Computing, pp. 392–399 (2016)

---

[4] https://www.openhab.org/docs/.

6. da Cruz, M.A.A., Rodrigues, J.J.P.C., Al-Muhtadi, J., Korotaev, V.V., de Albuquerque, V.H.C.: A reference model for internet of things middleware. IEEE Internet Things J. **5**(2), 871–883 (2018)
7. Fattah, S.M.M., Sung, N.M., Ahn, I.Y., Ryu, M., Yun, J.: Building IoT services for aging in place using standard-based IoT platforms and heterogeneous IoT products. Sensors **17**(10), 2311 (2017)
8. Guth, J., et al.: A detailed analysis of IoT platform architectures: concepts, similarities, and differences. In: Di Martino, B., Li, K.-C., Yang, L.T., Esposito, A. (eds.) Internet of Everything. IT, pp. 81–101. Springer, Singapore (2018). https://doi.org/10.1007/978-981-10-5861-5_4
9. Hamza, A.A., Abdel Halim, I.T., Sobh, M.A., Bahaa-Eldin, A.M.: HSAS-MD analyzer a hybrid security analysis system using model-checking technique and deep learning for malware detection in iot apps. Sensors **22**, 1079 (2022)
10. Hoang, T.S., Kuruma, H., Basin, D., Abrial, J.R.: Developing topology discovery in Event-B. Sci. Comput. Program. **74**(11–12), 879–899 (2009)
11. Houimli, M., Kahloul, L., Benaoun, S.: Formal specification, verification and evaluation of the MQTT protocol in the Internet of Things. In: 2017 International Conference on Mathematics and Information Technology (ICMIT), pp. 214–221 (2017)
12. Kim, J., Lee, J.: OpenIoT: an open service framework for the Internet of Things. In: 2014 IEEE World Forum on Internet of Things (WF-IoT), pp. 89–93. IEEE (2014)
13. Krishna, A., Le Pallec, M., Mateescu, R., Noirie, L., Salaün, G.: Rigorous design and deployment of IoT applications. In: Proceedings of the 7th International Workshop on Formal Methods in Software Engineering, FormaliSE@ICSE 2019, Montreal, QC, Canada, 27 May 2019, pp 21–30 (2019)
14. Muthukumar, N., Srinivasan, S., Ramkumar, K., Pal, D., Vain, J., Ramaswamy, S.: A model-based approach for design and verification of industrial internet of things. Future Gener. Comput. Syst. **95**, 354–363 (2019)
15. Patel, P., Cassou, D.: Enabling high-level application development for the Internet of Things. J. Syst. Softw. **103**, 62–84 (2015)
16. Salahuddin, M.A., Al-Fuqaha, A., Guizani, M., Shuaib, K., Sallabi, F.: Softwarization of IoT infrastructure for secure and smart healthcare. IEEE Comput. **50**(7), 74–79 (2017)
17. Salihbegovic, A., Eterovic, T., Kaljic, E., Ribic, S.: Design of a domain specific language and IDE for Internet of things applications. In: 2015 38th International Conference on Information and Communication Technology, Electronics and Microelectronics (MIPRO), pp, 996–1001 (2015)
18. Sherratt, E., Ober, I., Gaudin, E., Fonseca i Casas, P., Kristoffersen, F.: SDL - the IoT language. In: Fischer, J., Scheidgen, M., Schieferdecker, I., Reed, R. (eds.) SDL 2015. LNCS, vol. 9369, pp. 27–41. Springer, Cham (2015). https://doi.org/10.1007/978-3-319-24912-4_3
19. Sosa-Reyna, C.M., Tello-Leal, E., Alabazares, D.L.: Methodology for the model-driven development of service oriented IoT applications. J. Syst. Architect. - Embed. Syst. Des. **90**, 15–22 (2018)
20. Stevens, C., Alhanahnah, M., Yan, Q., Bagheri, H.: Comparing formal models of IoT app coordination analysis. In: ACM SIGSOFT WOrkshop on Software Security (SEAD'20), pp. 3–10. ACM (2020)
21. Thramboulidis, K., Christoulakis, F.: UML4IoT-A UML-based approach to exploit IoT in cyber-physical manufacturing systems. Comput. Ind. **82**, 259–272 (2016)

Differentiation Between Normal and Abnormal Functional Brain Connectivity Using Non-directed Model-Based Approach

Heba Ali[1]([✉])(iD), Mustafa A. Elattar[1,2](iD), Walid Al-Atabany[1,2](iD), and Sahar Selim[1,2](iD)

[1] Medical Imaging and Image Processing Research Group, Center for Informatics Science, Nile University, Sheikh Zayed City 12588, Egypt
h.ali2137@nu.edu.eg
[2] School of Information Technology and Computer Science, Sheikh Zayed City 12588, Egypt
https://nu.edu.eg/index.php

Abstract. Brain Connectivity refers to networks of functional and anatomical connections found throughout the brain. Multiple neural populations are connected by intricate connectivity circuits and interact with one another to exchange information, synchronize their activity, and participate in the accomplishment of complex cognitive tasks. Issues about how various brain regions contribute to cognition and their reciprocal roles have drawn the attention of researchers since the beginning of neuroscience. The interest in brain connection estimation has grown significantly due to the advancement of cutting-edge functional brain imaging techniques and the fine-tuning of sophisticated signal-processing algorithms. This study investigates the difference between functional brain connectivity patterns for normal and abnormal cases. The brain connectivity is quantified by encoding neighborhood relations into a connectivity matrix using a Non-directed Model-based technique, whose columns and rows represent various brain regions. This representation is then translated into a graphical model, which offers a way to quantify different topological data. The proposed approach enables us to study the pairwise relations between interacting brain regions. The results show that the normal subjects have the same functional connectivity pattern with 96% overall normal subjects. Moreover, most abnormal cases have different functional connectivity patterns than normal, depending on their abnormality.

Keywords: Brain connectivity pattern · Functional connectivity · Non-directed model-based · EEG

Supported by Nile University.

P. Fournier-Viger et al. (Eds.): MEDI 2022, CCIS 1751, pp. 89–102, 2022.
https://doi.org/10.1007/978-3-031-23119-3_7

1 Introduction

Understanding the neuronal activity that organizes human behavior is one of neuroscience's primary goals in clarifying this relationship. Various studies have concentrated on the alleged connections between task-related neuronal electrical activity and behavioral performance. Meanwhile, studies using BOLD (blood-oxygen-level-dependent) imaging have found a correlation between resting-state connectivity and functionally activated networks, particularly regarding the execution of different cognitive functions like working memory, intelligence, and attention. Additionally, it has been proposed that these correlations could be explained by the unique properties of the brain network [10,18]. Understanding the connectivity patterns of the human brain provides crucial details about the structural, functional, and causal organization of the brain. Functional and efficient connectivity methods for the brain have recently been the subject of computational studies [7]. The brain's functional organization is characterized by the integration and segregation of processed information. The idea that anatomical and functional connections between brain regions are set up in such a way as to facilitate information processing is a core pillar of emerging neuroscience. The synchronized activity provides functional connectivity locally and between dispersed brain regions. Thus, brain networks are made up of functionally connected but spatially dispersed areas that process information. Three distinct but related types of connectivity serve as the foundation for brain connectivity analysis [7]. Researchers are trying to use functional connectivity networks for treatment assessment for other brain disorder diseases because it is increasingly used as novel disease biomarkers or predictors of outcome and helps increase treatment performance [4,6].

The diagnosis of most brain diseases can be difficult at the onset or during presurgical evaluation, especially when the electroencephalography (EEG) does not show abnormalities. Moreover, behavioral impairments can be observed despite the absence of apparent abnormalities in EEG [5]. Brain Connectivity measures can investigate relationships between brain regions that belong to a network. EEG offers a more direct measure of neuronal activity and a higher temporal resolution that could be well suited to investigate dynamic brain processes compared to functional magnetic resonance imaging (fMRI) because they cannot provide information about the directionality of information flow [17].

In our study, Non-directed connectivity is measured, which shows the relation between the different regions in the brain without causal influence; We investigate it using the concept of correlation: it mainly measures the linear dependency between every pair of channels by establishing a connection between the two-time series by measuring the linear correlation [20]. This method has an advantage over the other techniques, which is creating networks with non-random topological characteristics. Networks that measure connectivity through correlation are, in particular, more naturally clustered than random networks [21]. Our study mainly aims to answer the following open question. Is there a proper match between connectivity patterns in normal individuals? And also, Is this pattern different from the abnormal individual's patterns?

This study serves two purposes: first, to gain a deeper understanding of the mechanisms underlying both healthy and pathological brain functional connectivity, and second, to inspire new and creative therapeutic approaches in the future. Assume we know the neural circuitry changes that occur in neurological disorders and the key brain areas involved. If so, these can then be the focus of treatments intended to return the body to normal function or, at the very least, to minimize the most severe symptoms. Brain connectivity may offer extremely perceptive neuro markers of brain disorders, which can detect subtle alterations brought on by the development or improvement of the disorder and can detect changes even at an early preclinical stage. This is a fascinating aspect in this regard that still warrants further research. Thus, connectivity-based neuro markers may also help forecast a disorder's clinical course or a therapeutic intervention's behavioral or motor effects.

We included in our paper the research methodology in the first section, Dataset and Preprocessing in the second section. Then Experimental Results and Discussion.

2   Methodology

This paper aims to investigate the differences between functional connectivity patterns of normal and abnormal people to use these findings to enhance the treatment performance for brain disorder diseases. The proposed methodology consists mainly of two components which are preprocessing and Brain Connectivity Network, in which the brain connectivity network includes functional connectivity and similarity measurement, as illustrated in Fig. 1.

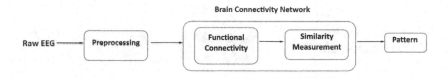

Fig. 1. Working process of the proposed methodology

2.1   Brain Connectivity Network

Brain Connectivity is the influence one brain region can have on another brain region. It is divided into three types [16].

The Anatomical (Structure) Connectivity. The structural (synaptic) connections link sets of neurons and their associated structural biophysical attributes encapsulated in parameters such as synaptic strength.

Functional Connectivity. It captures deviations from statistical independence between distributed and spatially remote neuronal units; It is estimated by calculating correlation or covariance, spectral coherence, and phase-locking.

Effective Connectivity. It describes networks of directional effects of one neural over another and unites structural and functional Connectivity; causal effects can be inferred through methodical perturbations of the system because causes must come before effects in time through time series analysis [16].

In our study, we focused on measuring the functional Connectivity among normal and abnormal people linearly using a Non-Directed Model-based approach to investigate the differences between the pattern of normal and abnormal ones.

2.2   Functional Connectivity

Functional Connectivity can be measured differently, as shown in Fig. 2. A first subdivision Based on the metric, the direction of the interaction. Non-directed functional connectivity metric quantify some degree of signal interdependence without considering the influence's directional nature. Additionally, the directed methods aim to prove statistical causation using the data and their results. [9]. Another division between model-free and model-based approaches for both directed and non-directed types. The model-based approaches used information theory concepts to quantify the generalized (linear and non-linear) interdependence between two or more variables (or time series). All assume linearity concerning the types of interactions between two signals. Finally, model-free methods for detecting directed interactions [13]. The model-free methods help quantify non-linear neuronal interactions. Linear methods are sufficient to capture various oscillatory interactions [3].

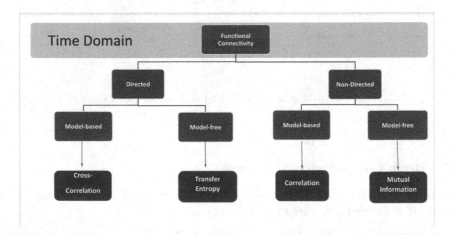

Fig. 2. Popular methods for measuring functional connectivity in time domain

2.3 Pearson's Correlation

In our study, we measured Functional Connectivity using a Non-directed Model-based approach; we used Pearson's Correlation, which measures the linear dependence between two different signals for two-time series, X and Y. The key component of the framework of Pearson correlation is the correlation coefficient (COR), which measures the percentage of the variance of one of the signals that can be explained by the other and vice versa [8, 19].

Let Xi (t) and Yj (t) denote, respectively, the EEG signals from electrodes i and j, and where $\overline{X}$ and $\overline{Y}$ are the means of the Xi and the Yj, respectively, and the summations $\sum$ are each overall data.

By calculating the linear correlation between the two time series in accordance with Eq. 1, COR establishes a connection between them. The range of it is $[1, -1]$. The two boundaries between the two-time series are $1/-1$ for perfect linear positive/negative correlation and 0 for no linear dependency.

$$Pcorxy\& = \left(\frac{\sum(X_i - \overline{X})(Y_j - \overline{Y})}{\sqrt{\sum(X_i - \overline{X})^2 \sum(Y_j - \overline{Y})^2}} \right) \tag{1}$$

2.4 Similarity Measurement of Functional Connectivity Matrix

In this section, we measured the similarity of the Functional Connectivity matrix among all subjects to assess the structural relationships among subjects commonly used across disciplines in neuroscience [22]. Then we used the rank correlation [1] as a metric to measure the similarity, as shown in Eq. 2. The Spearman's ρ is used because it doesn't assume that similarity increases linearly. Where di represents the difference in paired ranks and n represents the number of cases.

$$\rho xy\& = \left(1 - \frac{6 \sum(di)^2}{n((n)^2 - 1)} \right) \tag{2}$$

Spearman's rank coefficient assesses the monotonic relationship between two Functional Connectivity Matrix. We first compute the ranks of the sample values, then replace them with ranks. Where the smallest value is replaced with 1, the next smallest is replaced with 2, and so on [23].

2.5 Similarity Analysis

Brain Connectivity research is naturally based on finding the relationship between different channels and brain regions. Clinicians gather information to predict clinical diagnosis and treatment. Therefore, there is an endless struggle to connect what is already known to what needs to be known. In our study, we are trying to infer the Functional Connectivity pattern similarity for normal people compared to abnormal people. The most basic mathematical form used for such analysis is correlational analysis [23]. This section presents a statistical analysis of the Functional Connectivity Matrix Similarity. Since Spearman's rho usually have values between $-1 \leq \rho xy \leq 1$.

A positive Spearman is interpreted as a perfect monotonically growing relationship. A similar application for negative values [23]. We validate our results based on similarity analysis through the statistical test.

All statistical tests have a null hypothesis; our null hypothesis is that there is no relationship between the Functional Connectivity Matrix for the normal group and no differences among the abnormal group. The smaller the p-value associated with the Spearman correlation coefficients, the more likely you are to reject the null hypothesis. P-value is considered a number calculated from a statistical test. In hypothesis testing, the p-value is used to determine whether to reject the null hypothesis or not [14].

3   Data-Set and Preprocessing

The dataset used in this study is The NMT (NUST-MH-TUKL EEG) Scalp EEG Dataset: An open-source dataset of pathological and healthy EEG recordings [12]. In this section, we discuss the preparation and preprocessing steps.

3.1   Data Preparation

The dataset consists of 2,417 recordings (2002 normal and 415 abnormal) corresponding to 1,608 males, 808 females, and one unknown gender from unique participants spanning around 625 h. A team of qualified neurologists labeled each recording as normal or abnormal. The patient's age and gender are also included in the demographic data. The South Asian population is the main subject of the dataset. At the Military Hospital in Rawalpindi, a team of two qualified neurologists labeled data with the assistance of a technician [12].

Figure 3 shows the standard 10–20 system-based EEG montage used for recordings. Channels A1 and A2 on the auricle of the ear serve as references among the 19 channels on the scalp. All channels are sampled 200 Hz. Each record lasts, on average, 15 min. The records are collected from males and females, ranging from under one year old to 90 years old. In contrast to women, 16.17% of male EEG recordings are abnormal or pathological [12]. In our study, we randomly selected 64 subjects as a subset of the total subjects, 34 from the normal group and 34 from the abnormal group, to perform our study.

3.2   Data Preprossessing

We can extract meaningful information and further monitor a patient's health, diagnose, and identify various brain conditions from the complex biological electricity signal, which reflects the functional state of the brain connected to the

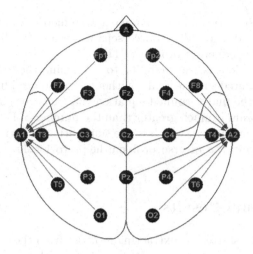

Fig. 3. The standard 10–20 system-based EEG montage used for recordings. Channels A1 and A2 on the auricle of the ear serve as references among the 19 channels on the scalp [12].

individual's mental condition. But because EEG has such a high temporal resolution, it is usually contaminated by various unwanted artifacts. The cause of artifacts is due to measurement instruments and human subjects, which may be faulty electrodes, line noise, and high electrode impedance [11]. Physiological artifacts, however, are more challenging to get rid of. Therefore, We have to perform preprocessing to detect and extract clean EEG data accurately. Figure 4 shows the pipeline for preprocessing the EEG data.

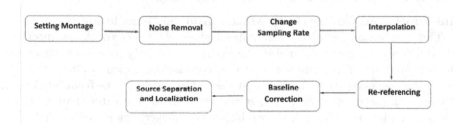

Fig. 4. Preprocessing pipeline

The preprocessing pipeline starts with setting the montage that contains the EEG sensor's position in 3D so it can be assigned to our current EEG data. It has a significant role in estimating forward and inverse solutions [11]. Applying the suitable filter help in removing noise and data cleaning, and then the interpolation step is to check pad channels and remove them after that baseline correction which helps to minimize the impact of temporal drifts that can be brought on by various internal and external factors and can vary over time and between electrodes. Finally, source separation and localization are critical due to the volume conduction problem [11].

The inverse problem is the inability to determine the active brain regions from a signal measured on the scalp without imposing additional restrictions [2]. This is due to the undetermined equation that must be solved. The number of neurons in the brain is much greater than the number of EEG electrodes; It is known as volume conduction, which is the summation of currents from different brain sources; solving the inverse problem helps to localize EEG responses in basic research [2].

4   Experimental Results

Our experiment was started by extracting a subset from the dataset, 34 subjects from the normal group and another 34 subjects from the abnormal group. Then go through the preprocessing pipeline in Fig. 3. Setting the montage on the 10–20 standards, applying a band-pass filter (1:35 Hz) to remove artifacts and clean the data; the sampling rate of the data 200 Hz. Then interpolating the data to remove the pad channels, the data was referenced to A1 and A2. Then implement the baseline correction by calculating the average voltage values for each electrode over a while. This average is then subtracted from the signal's time. Finally, we implement source separation and localization using Independent Component Analysis (ICA) [15]. We get 19 independent components from applying the (ICA). The sensor data of each subject was then subjected to the unmixing matrix obtained from all subjects. After preprocessing the data, the functional connectivity is studied using a Non-directed Model-based approach between each pair of EEG channels, then construct a matrix that represents the Functional Connectivity as shown in Fig. 5 for the normal subjects.

The Functional Connectivity Matrix Analysis is as follows:
    The correlation between each pair of channels refers to the connectivity between this pair, every channel in the matrix is perfectly correlated with itself. The stuff on the off diagonals is called R correlation, which is about the self-correlation between each channel and itself. Our scale starts from -0.6 to 1, meaning complete correlation at 1, zero correlation at 0, greater than 0.5 is significant, and less than 0.5 is less correlation; the darker regions tell us that there is a network and the lighter regions tell us that there is a strong network. It also looks like there's even some sub-grouping, which means that a group of channels has a great network among them. The matrix follows the anatomy of the brain well.

    Then the Spearman coefficient is used to check the similarity of the functional connectivity matrix between each pair of subjects. We measured the similarity between every two subjects by extracting the top triangle from the matrix as shown in Fig. 6. It offers more consistency with other matrix functions while

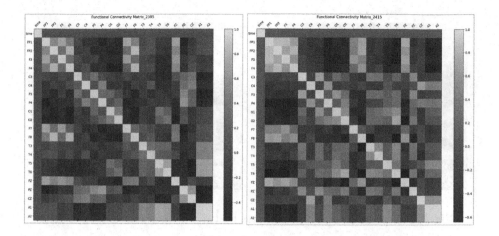

Fig. 5. Functional Connectivity Matrix for two different normal subjects

recasting the upper triangle into a matrix. Furthermore, the p-value is used beside the Spearman coefficient to validate our results in this stage.

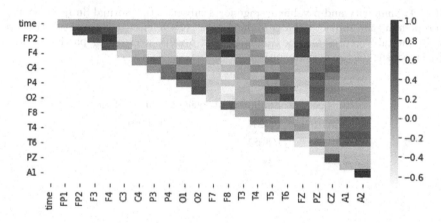

Fig. 6. The upper triangle of Functional Connectivity Matrix

We first compare every normal subject with all normal subjects, then take the average of the p-value and the Spearman coefficient representing the amount of similarity among normal subjects, as illustrated in Table 1 The Similarity Average and p-value in Table 1 indicates that there is significant similarity among the normal subjects. Then the abnormal subjects are studied together and compared to each other, as shown in Table 2. This indicates a very weak similarity among the abnormal subjects because they have different brain diseases and severity.

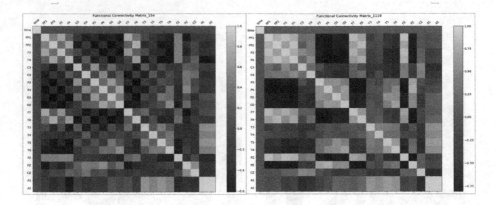

Fig. 7. Functional Connectivity Matrix for two different abnormal subjects

And the normal subjects are compared with the abnormal subjects, as shown in Table 3. In every iteration, the functional connectivity is measured for a pair of subjects and compared in their similarity analysis. That gives us intuition about a very weak similarity between normal and abnormal subjects.

Table 1. Similarity and p-value average for a subset of the Normal Subjects. (Subject ID) represents the subject number in the dataset; (N) refers to a Normal subject; Every row in the table represents the Similarity and p-value average for particular normal subjects versus all the normal subjects

Normal subjects		
Subject ID	Similarity average	Pvalue average
N2410	0.9234	<0.00001
N1426	0.8087	<0.0001
N537	0.9648	<0.00001

5   Discussion

The functional connectivity patterns are investigated using a Non-directed Model-based approach on the NMT (NUST-MH-TUKL EEG) Scalp EEG Dataset. Our findings will improve the treatment assessment as functional connectivity patterns are used as novel disease biomarkers and have become a focus for new treatment strategies that aim to reduce neurological deficits. The differentiation between normal and abnormal patterns will enhance treatment performance for neural disorder diseases.

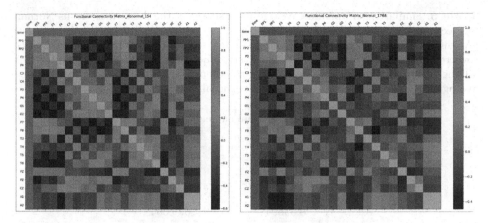

Fig. 8. Functional Connectivity Matrix for two different normal and abnormal subjects

Table 2. Similarity and p-value average for a subset of the Abnormal Subjects. (Subject ID) represents the subject number in the dataset; (A) refers to an Abnormal subject; Every row in the table represents the Similarity and p-value average for a particular abnormal subject versus all the abnormal subjects

Abnormal subjects		
Subject ID	Similarity average	p-value average
A557	0.5346	<0.00001
A2264	0.337	<0.000001
A18	0.2029	<0.000001

The results show that the normal subjects have the same pattern with 96% accuracy; overall, the normal subjects that have been used from the dataset. Meanwhile, abnormal subjects have different patterns depending on their kind of abnormality and its severity. Figure 7 shows different patterns among the abnormal subjects. While investigating the normal subjects versus the abnormal subjects, we find a significant difference between them, as shown in Fig. 8. During validating our observation by checking the Functional Connectivity Matrix Similarity for each group, the similarity among normal subjects is (correlation = 0.8 ± 0.2809) and (p-value <0.00001) as illustrated in Table 1, which is a good indicator of the significance of our findings. For abnormal subjects, the similarity among abnormal subjects is (correlation = 0.4 ± 0.1997) and (p-value <0.0001) meanwhile comparing the normal subjects versus the abnormal subjects, it is a very poor correlation among the Functional Connectivity Matrix (correlation = 0.2 ± 0.1872) and (p-value <0.000001) which validate our findings as shown in Table 3.

In conclusion, our findings show a Functional Connectivity similarity among the normal subjects; meanwhile, each abnormal subject has a different Functional

Table 3. Similarity and p-value average for a subset of the Abnormal versus Normal Subjects and vice versa. (Subject ID) represents the subject number in the dataset; (A) refers to an Abnormal subject, and (N) refers to a Normal subject; The first three rows represent the Similarity and p-value average for a particular normal subject versus all the abnormal subjects; the next three rows represents the Similarity and p-value average for a particular abnormal subject versus all the normal subjects

Normal versus Abnormal subjects		
Subject ID	Similarity average	p-value average
N2410—A	0.3965	<0.00000003
N1426—A	0.50603	<0.00001
N537—A	0.20359	<0.0000001
A557—N	0.53694	<0.00004
A2264—N	0.45319	0.0004
A18—N	0.23475	0.000319

Connectivity pattern depending on the kind and severity of the abnormality. We plan to extend our study to be based on a Directed Model-based approach to allow us to investigate more features of brain connectivity fingerprints for each subject with a Non-linearly approach.

Conclusion

Functional Connectivity pattern is increasingly used as novel disease biomarkers or predictors of outcome. Also, it has become a focus for new treatment strategies aiming at enhancing performance or just reducing neurological deficits. These revelations might open new avenues for training and learning in clinical practice and public usage. Our study illustrates the differences between Functional Connectivity patterns for normal and abnormal people by constructing a Functional Connectivity Matrix from low-dimensionality (low-density) EEG using a Non-directed Model-based approach, considered common practice to compute sensor Functional Connectivity among electrodes linearly. This study shows that all the normal cases used from the NMT (NUST-MH-TUKL EEG) Scalp EEG Dataset shared the same Functional Connectivity pattern, with 96 % overall normal subjects. In contrast, all the abnormal cases have different Functional Connectivity patterns based on their abnormality. Our findings are built on the correlation analysis for the Functional Connectivity Matrix, which is a non-directed model-based. However, this approach has an advantage over other techniques for creating networks with non-random topological characteristics. And the Networks that measure connectivity through correlation are, in particular, more naturally clustered than random networks; non-linear methods can offer more features and characteristics. Our future work will consider non-linear methods, integrate them, and compare them with the current proposed approach for better assessment.

References

1. Akoglu, H.: User's guide to correlation coefficients. Turkish J. Emerg. Med. **18**(3), 91–93 (2018)
2. Awan, F.G., Saleem, O., Kiran, A.: Recent trends and advances in solving the inverse problem for EEG source localization. Inverse Prob. Sci. Eng. **27**(11), 1521–1536 (2019)
3. Bastos, A.M., Schoffelen, J.M.: A tutorial review of functional connectivity analysis methods and their interpretational pitfalls. Front. Syst. Neurosci. **9**, 175 (2016)
4. Chin Fatt, C.R., et al.: Effect of intrinsic patterns of functional brain connectivity in moderating antidepressant treatment response in major depression. Am. J. Psychiatry **177**(2), 143–154 (2020)
5. Coito, A., et al.: Altered directed functional connectivity in temporal lobe epilepsy in the absence of interictal spikes: a high density EEG study. Epilepsia **57**(3), 402–411 (2016)
6. Finn, E.S., Constable, R.T.: Individual variation in functional brain connectivity: implications for personalized approaches to psychiatric disease. Dialogues Clin. Neurosci. (2022)
7. Finn, E.S., et al.: Functional connectome fingerprinting: identifying individuals using patterns of brain connectivity. Nat. Neurosci. **18**(11), 1664–1671 (2015)
8. Geerligs, L., Henson, R.N., et al.: Functional connectivity and structural covariance between regions of interest can be measured more accurately using multivariate distance correlation. Neuroimage **135**, 16–31 (2016)
9. Granger, C.W.: Investigating causal relations by econometric models and cross-spectral methods. Econometrica: J. Econometric Soc. 424–438 (1969)
10. Hermundstad, A.M., et al.: Structural foundations of resting-state and task-based functional connectivity in the human brain. Proc. Natl. Acad. Sci. **110**(15), 6169–6174 (2013)
11. Jiang, X., Bian, G.B., Tian, Z.: Removal of artifacts from EEG signals: a review. Sensors **19**(5), 987 (2019)
12. Khan, H.A., et al.: The NMT scalp EEG dataset: an open-source annotated dataset of healthy and pathological EEG recordings for predictive modeling. Front. Neurosci. **15**, 1764 (2021)
13. Lobier, M., Siebenhühner, F., Palva, S., Palva, J.M.: Phase transfer entropy: a novel phase-based measure for directed connectivity in networks coupled by oscillatory interactions. Neuroimage **85**, 853–872 (2014)
14. Mark, D.B., Lee, K.L., Harrell, F.E.: Understanding the role of p values and hypothesis tests in clinical research. JAMA Cardiol. **1**(9), 1048–1054 (2016)
15. Pion-Tonachini, L., Hsu, S.H., Makeig, S., Jung, T.P., Cauwenberghs, G.: Real-time EEG source-mapping toolbox (rest): online ICA and source localization. In: 2015 37th Annual International Conference of the IEEE Engineering in Medicine and Biology Society (EMBC), pp. 4114–4117. IEEE (2015)
16. Rubinov, M., Sporns, O.: Complex network measures of brain connectivity: uses and interpretations. Neuroimage **52**(3), 1059–1069 (2010)
17. da Silva, F.L.: EEG and MEG: relevance to neuroscience. Neuron **80**(5), 1112–1128 (2013)
18. Stampanoni Bassi, M., Iezzi, E., Gilio, L., Centonze, D., Buttari, F.: Synaptic plasticity shapes brain connectivity: implications for network topology. Int. J. Mol. Sci. **20**(24), 6193 (2019)

19. Šverko, Z., Vrankić, M., Vlahinić, S., Rogelj, P.: Complex Pearson correlation coefficient for EEG connectivity analysis. Sensors **22**(4), 1477 (2022)
20. Wang, M., Hu, J., Abbass, H.A.: Brainprint: EEG biometric identification based on analyzing brain connectivity graphs. Pattern Recogn. **105**, 107381 (2020)
21. Zalesky, A., Fornito, A., Bullmore, E.: On the use of correlation as a measure of network connectivity. Neuroimage **60**(4), 2096–2106 (2012)
22. Zar, J.H.: Spearman Rank Correlation: Overview. Wiley StatsRef: Statistics Reference Online (2014)
23. Zhang, W.Y., Wei, Z.W., Wang, B.H., Han, X.P.: Measuring mixing patterns in complex networks by spearman rank correlation coefficient. Phys. A **451**, 440–450 (2016)

Database Systems

Benchmarking Hashing Algorithms for Load Balancing in a Distributed Database Environment

Alexander Slesarev[1,2], Mikhail Mikhailov[1], and George Chernishev[1,2(✉)]

[1] Unidata, Saint-Petersburg, Russia
alexandr.slesarev@unidata-platform.org,
{mikhail.mikhailov,georgii.chernyshev}@unidata-platform.ru
[2] Saint-Petersburg State University, Saint-Petersburg, Russia

Abstract. Modern high load applications store data using multiple database instances. Such an architecture requires data consistency, and it is important to ensure even distribution of data among nodes. Load balancing is used to achieve these goals.

Hashing is the backbone of virtually all load balancing systems. Since the introduction of classic Consistent Hashing, many algorithms have been devised for this purpose.

One of the purposes of the load balancer is to ensure storage cluster scalability. It is crucial for the performance of the whole system to transfer as few data records as possible during node addition or removal. The load balancer hashing algorithm has the greatest impact on this process.

In this paper we experimentally evaluate several hashing algorithms used for load balancing, conducting both simulated and real system experiments. To evaluate algorithm performance, we have developed a benchmark suite based on Unidata MDM—a scalable toolkit for various Master Data Management (MDM) applications. For assessment, we have employed three criteria—uniformity of the produced distribution, the number of moved records, and computation speed. Following the results of our experiments, we have created a table, in which each algorithm is given an assessment according to the abovementioned criteria.

Keywords: Consistent hashing · Databases · Benchmarking

1 Introduction

As any organization grows, the volume of its corporate data assets rises as well. There are two general approaches to solving this architectural problem [22]: vertical and horizontal scaling. Vertical scaling focuses on increasing the capabilities of a single server, whereas horizontal scaling involves adding machines to the cluster. To implement horizontal scaling, a database table has to be horizontally split into parts (shards), which are stored on different server nodes.

Horizontal scaling has several significant advantages, such as the possibility to flexibly adjust storage volume by altering cluster size. Another one is the ability to deal with data loss by replicating data among servers. Thus, the need for distributed data storage appears.

An important component of distributed storage is the load balancer—a mechanism which determines which particular server will store a data entity (e.g. record, table part, etc.). There are several assessment criteria for load balancers. First of all, data distribution over servers should be as close to uniform as possible. Next, if cluster size changes, the number of moved data entities must be close to optimal. And finally, load balancer computing costs should not be high.

In order to calculate the shard assigned to a given data entity, the load balancer utilizes a hashing algorithm. Since the 90's, many hashing algorithms have been designed specifically for balancing different types of loads such as network connection management, distributed computing optimization, and data storage balancing.

One of the research disciplines focusing on storage and processing of large data volumes is Master Data Management [1,19] (MDM). It is based around the concept of Master Data—a concept that combines objects important for business operations within an organization, such as inventory, customers, and employees. The main goals of MDM are unification, reconciliation, and ensuring completeness of corporate Master Data.

In this paper, we compare several hashing algorithms and assess their applicability to the load balancing problem. We experimentally evaluate them using both simulated and real tests. For the latter, we employ the Unidata platform [10]—an open-source MDM toolkit that has distributed storage capabilities. Following the results of our experiments, we have created a table, in which each algorithm is given an assessment according to the abovementioned criteria.

This paper is organized as follows. In Sect. 2 we describe a number of existing load balancing algorithms, define several terms from the MDM area, and review Unidata storage architecture. Then in Sect. 3 we describe the conducted experiments, and discuss the achieved results in Sect. 4. We conclude this paper with Sect. 5.

2   Background and Related Work

In this section we describe those existing hashing algorithms that we are going to benchmark, and since our last series of experiments is run on a real system, we also provide a general description of the system itself, its purpose, and the used data schema.

This study concerns two research fields that have a rich body of work—load balancing and hashing algorithms. The former has a large number of surveys describing dozens of works. For example, consider study [8], which references many more similar surveys. Unfortunately, such surveys only classify the covered methods using some high-level criteria (e.g. adaptivity, static or dynamic, preemptiveness, etc.). They do not experimentally evaluate surveyed algorithms.

The reason for this is the fact that such surveys are too broad, their reviewed studies belong to many different fields and it will be extremely difficult to run such an evaluation. At the same time, industry is interested in the best method for a particular domain, and the answer can be found only empirically.

Turning to hashing, we must mention a very comprehensive survey [3], which describes many hashing approaches, and proposes an algorithm taxonomy. However, the section that concerns data-oriented hashing is aimed towards data structures and machine learning, but not load balancing. The set of hashing algorithms that we consider in our study is absent in this survey.

Therefore, our work fills the gap in existing studies.

2.1 Considered Methods

Let us start with the hashing algorithms which are used for data balancing. Each of the considered load balancers applies its hash function to some incoming data entity, so we call the result of the application a *key*. The purpose of a load balancer is to match each key to one of the shards, which is represented by an integer number (id). There are several hashing methods that we consider in this paper:

- **Linear Hashing** is one of the overall oldest algorithms. Besides the classic version [2] there is a number of modifications such as LH* [18], LH*M [14], LH*G [16], LH*S [15], LH*SA [13] and LH*RS [17]. The core idea of algorithms of this family is to calculate the remainder of dividing the key by a number of shards in the system. Therefore, they are suitable for solving the problem featuring a fixed number of shards, which in our case is a disadvantage. In our work we have adopted a version used for partitioning in PostgreSQL[1].
- **Consistent Hashing.** Originally, this algorithm [9] was designed for balancing loads of computer networks. Nowadays, it seems to be the most popular method for balancing many various types of loads. For example, distributed systems like AWS DynamoDB [4] and Cassandra [11] use Consistent Hashing for partitioning and replication. This method is based on picking random points on a ring, which is a looped segment of real numbers that represents shards and data entities as points. Points denoting keys are assigned to the clockwise nearest shard. To ensure even data distribution, each shard is represented by several points. Note that the design of this method allows to change the number of shards while moving only the optimal number of records.
- **Rendezvous.** Similarly to Consistent Hashing, this method [21] was also developed to optimize network load. For a given key, the algorithm calculates the value of the cost function for each of the shards and assigns the key to the shard with the highest value. When adding or removing shards, Rendezvous also does not move extra records.

---

[1] https://github.com/postgres/postgres/blob/master/src/backend/partitioning/
partbounds.c.

- **RUSH** [6] was developed to store data in a disk cluster. It has two modifications: $RUSH_R$ and $RUSH_T$ [7]. Authors of RUSH focused on improving the uniformity of data distribution when changing the size of the cluster, therefore, the algorithm is based on the following principle: every time cluster size is changed, a special function is used to decide which objects should be moved to balance the system.

- **Maglev** is an algorithm [5] from Google's load balancer for web services. The goal of Maglev is to improve data uniformity (compared to Consistent Hashing) and cause "minimal disruption", e.g. if the set of shards changes, data records will likely be sent to the same shard where they were in before. It was proposed as a new type of Consistent Hashing, in which the ring is replaced with a lookup table by which a key can be assigned to a shard. The size of the lookup table should be greater than the possible number of shards to decrease collision rate. Average proposed lookup time is $O(Mlog(M))$, where M is size of lookup table.

- **Jump** is another load balancer from Google [12]. Its authors presented it as a superior version of Consistent Hashing which "requires no storage, is faster, and does a better job of evenly dividing the key space among the buckets". Jump generates values for shard numbers only in the $[0; \#shards]$ range, so the addition of a new shard is fast. However, a deletion of an intermediate shard will cause rehashing of many records. It takes $O(log(N))$ time to run Jump, where N is the number of shards.

- **AnchorHash.** According to the authors, AnchorHash [20] is a "hashing technique that guarantees minimal disruption, balance, high lookup rate, low memory footprint, and fast update time after resource additions and removals". A notable difference of AnchorHash from other discussed algorithms is that it stores some information about previous states of the system.

Each paper that proposed a novel hashing algorithm compared it only to a small number of other such algorithms. To the best of our knowledge, there were no dedicated comparisons of such algorithms applied to the horizontal scaling problem. At the same time, ensuring high performance of horizontal scaling is a pressing problem that is in demand by the industry. Thus, there is a need to evaluate all of these algorithms and study their applicability to this problem.

2.2 Basic Definitions

To understand the specifics of the data storage in which the load balancers will be evaluated, it is necessary to introduce some MDM terms:

- **Golden Record.** One of the main problems in MDM is the compilation and maintenance of a "single version of the truth" [1] for a given entity, e.g. person, company, order, etc. To achieve such a goal, an MDM system has to assemble information from many data sources (information systems of a particular organization) into one clean and consistent entity called the golden record.

– **Validity Period** is a time interval in which the information about an entity is valid. For each golden record, several validity periods may exist. This fact should be taken into account while querying the data. There are two temporal dimensions: time of an event and time of introduction of this new version of information to the system. This leads to a special storage scheme for managing this information.

2.3 System Architecture

MDM systems are a special class of information management systems [10]. Their specifics impose requirements on the platform storage architecture and data processing.

First of all, versioning of stored objects must be supported. For this reason, data assets describing stored objects have validity periods and they should be taken into account while querying the data.

Second, deletion operations can be performed only by an administrator, while a user can only mark data entity as removed. This is necessary to avoid information loss and ensure correct versioning support. Sometimes there are legal requirements for this data handling semantics. Such an architectural pattern is often called tombstone delete.

Third, provenance should be provided. It means that any system operation should be traceable. For example, there must be a way to roll back all changes in records after each operation.

The proposed approach is based on the following four tables, where three of them represent entities:

– Etalon stores the metadata of the golden record itself.
– Origin stores the metadata related to the source system of the record.
– Vistory (version history) is the validity period of origin, which in turn may have revisions.
– External Key is a table needed for accessing the data from within other parts of the Unidata storage.

The relations between these tables are shown in Fig. 1. The links with empty arrowheads denote "shared" (inherited) attributes and full arrowheads show the PK-FK relationship. The detailed descriptions of table attributes can be found in [10].

3 Evaluation

In order to select the best hashing algorithm for the load balancing problem, we have performed an experimental evaluation.

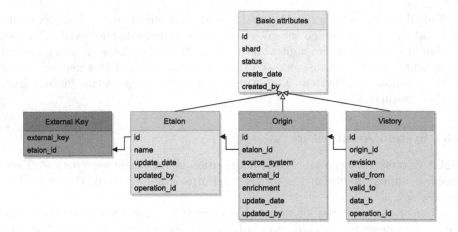

Fig. 1. Tables used for data storage in the Unidata platform

3.1 Experimental Setup

Experiments were performed using the following hardware and software configuration:

- Hardware: LENOVO E15, 16GiB RAM, Intel(R) Core(TM) i7-10510U CPU @ 4.90 GHz, TOSHIBA 238GiB KBG40ZNT.
- Software: Ubuntu 20.04.4 LTS, Postgres 11.x, JDK 11.x, Tomcat 7.x, Elasticsearch 7.6.x.

Some algorithms have parameters that affect their performance:

- For Consistent Hashing, 16 points for each shard were selected. This number was chosen experimentally, as a compromise between the hashing speed and uniformity of the initial distribution.
- For Maglev, lookup table size was set to 103. Similarly to Consistent Hashing, this number was selected experimentally. Note that this value is important in rebalancing process, but not for lookup.
- For AnchorHash, we have set the $|\mathcal{A}|$ (the number of buckets that algorithm works with) to double the maximum number of shards (i.e. 64) as it was recommended in the original paper [20].

3.2 Results

In order to evaluate the load balancing algorithms, we have defined three criteria which we ranked by their importance (in descending order):

1. Uniformity of the produced data distribution.
2. Redundant movement of records during shard addition or removal.
3. Lookup speed.

To select the best algorithm, we have conducted three experiments. First, we have performed a load balancing simulation experiment in Google Colab[2] (in Python). This step was needed to run a shallow, preliminary assessment of algorithm performance. It was conducted as follows: first, 10K records were generated and distributed (via hashing function) into 32 shards. Thus, uniform distribution will result in 312 records per shard. After this, 8 shards were scheduled for removal and system was forced to rebalance the data. Therefore, uniform distribution should result in 416 records per shard. This procedure was run for each of the considered load balancers (hashing functions).

The mean values of ten such experiments are presented in Table 1. The first column of the table contains the average time of shard calculation, the second and the third present variance of records assigned to shards before and after rebalancing, respectively. The last column shows the ratio of the number of actually moved records to the optimal number.

Table 1. First experiment, simulation in Colab

Algorithm	Shard id calculation time (ns)	Variance before drop	Variance after drop	Moved records ratio
Consistent	55049	72	94	1.00
Rendezvous	105331	18	21	1.00
RUSH$_R$	547044	95	125	1.57
Maglev	1146	16	21	1.39
Jump	18077	21	29	3.63
AnchorHash	3539	17	20	1.00

Based on the experimental results, we have decided to exclude RUSH$_R$ from further consideration due to it failing to conform to all three criteria. We have also excluded Jump due to the poor quality of rebalancing.

Our next experiments involve the Unidata platform, which is implemented in Java. To verify the consistency and transferability of previously obtained results, we have decided to re-evaluate the shard id calculation time inside the platform. Therefore, the second experiment was also to distribute 10K records among shards. The measured averages were as follows:

- Linear Hashing—808 ns
- Consistent Hashing—2419 ns
- Rendezvous—5945 ns
- Maglev—807 ns
- AnchorHash—2015 ns

---

[2] https://colab.research.google.com/drive/1pbJUFFP9JsSTSn7nrWv0tYdiUg2uRxA
   v?usp=sharing.

As one can see, the order of algorithm run times has not changed compared to the previous experiment. Therefore, we can conclude that switching the programming language did not affected the results of the previous experiment and we can continue using the Unidata platform.

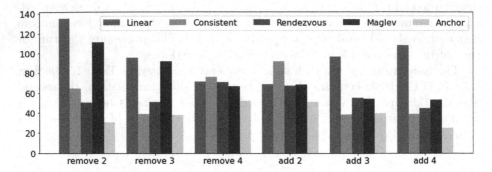

Fig. 2. Rebalance time (seconds).

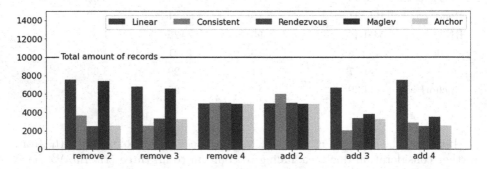

Fig. 3. Number of records from the Etalon table that were moved in the rebalancing process.

The third experiment was performed using a deployed Unidata platform. Its storage configuration was the following: four Docker nodes with PostgreSQL with eight shards on each node. We have generated 10K external keys and etalons as a workload. The idea of the experiment was as follows: remove three nodes one by one and then add them back in a similar manner.

The results of the evaluation are displayed in the following figures. Total time spent on each rebalance step is shown in Fig. 2, and the number of moved etalons is shown in Fig. 3. We have omitted such figure for external keys since it is largely the same (it is 1:1 mapping). Data distribution among shards on each step is shown in Figs. 4 and 5.

This experiment allows us to draw the following conclusions:

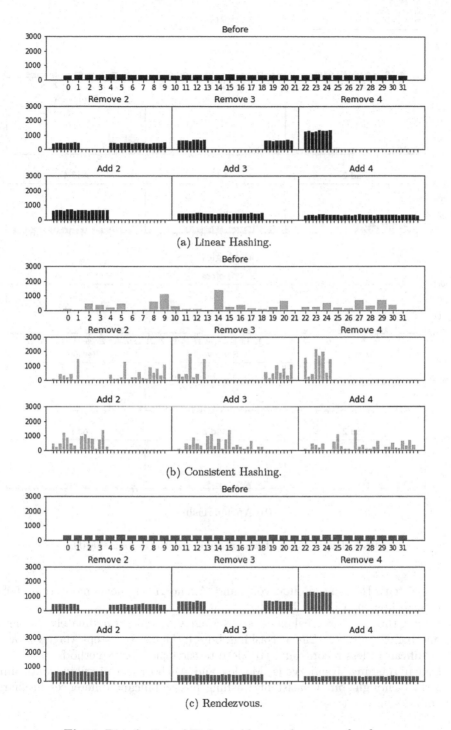

Fig. 4. Distribution of Etalon table records among shards.

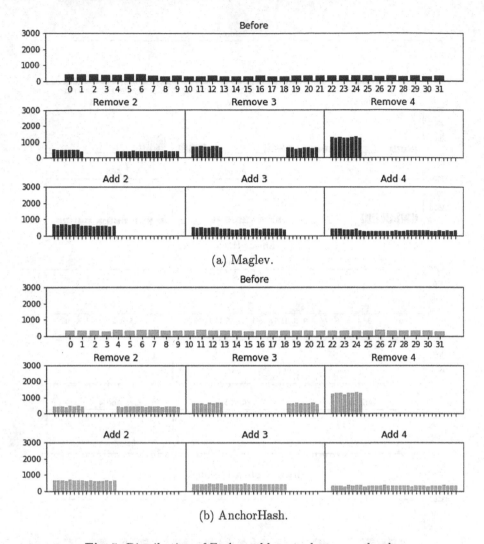

(a) Maglev.

(b) AnchorHash.

Fig. 5. Distribution of Etalon table records among shards.

- Consistent Hashing, Rendezvous and AnchorHash move more than 50% records less than Linear Hashing.
- During the first two rebalancing steps Maglev moved approximately the same number of records as Linear Hashing, but on the last two steps Maglev moved significantly less records and got close to the other three methods.
- Linear Hashing, Rendezvous, Maglev and AnchorHash distribute data uniformly enough, but Consistent Hashing has significant differences in shard volumes.

4 Discussion

Let us now discuss the compliance of each of the considered algorithms with the criteria defined in Sect. 3.

- **Linear Hashing** has proper record distribution among shards and high lookup speed, but it moves up to 80% of records on each rebalancing step, so this method does not meet our criteria. However, Linear Hashing can be applied in systems with a constant number of shards.
- **Consistent Hashing** has acceptable lookup times while moving an optimal number of records, but distributes data extremely unevenly. Methods with more uniform distribution should be preferred. To improve the distribution quality, one can increase the number of points for each shard on the ring, but this will slow down lookups.
- **Rendezvous** is optimal in case of data rebalancing and distribution, but has the longest lookup time. Since lookup speed is the least prioritized criterion, this method is suitable for us.
- **RUSH_R** does not satisfy all three criteria, so it does not suit our goals.
- **Maglev** provides fast lookup and relatively uniform distribution, but in some cases it can move more than 50% of all records (see the horizontal line on Fig. 3). Therefore, Maglev is appropriate for systems with a fixed number of shards.
- **Jump** moved the largest number of records (Table 1), so it is unsuitable as well.
- **AnchorHash** appears to be the winner so far as it satisfies all the requirements.

Following the results of all three experiments, we have created a table where we listed all evaluated algorithms (Table 2). We have assessed them according to our three criteria and assigned a rating out of three quality grades—low, medium, and high.

It is evident from the table that there are two winning algorithms—Maglev and AnchorHash, which fail to reach either top rebalancing quality (number of moved records) or the top lookup speed.

AnchorHash distributes data uniformly, moves an optimal amount of records and its lookup time is small enough. Rendezvous also fits first and second criteria, but its lookup time is more than two times larger than that of AnchorHash. These two methods are appropriate for systems with frequent shard addition or removal.

On the other hand, Maglev's lookup is more than two times faster, therefore it is suitable for static systems, similarly to Jump and Linear Hashing.

Consistent Hashing seems to be effective for both types of systems, but its main drawback is non-uniform data distribution among shards.

RUSH_R has been proven to be the worst algorithm out of all.

Table 2. Load balancer criteria satisfaction table

Algorithm	Data distribution uniformity	Rebalancing quality	Lookup speed
Linear	High	Low	High
Consistent	Low	High	Medium
Rendezvous	High	High	Low
RUSH	Low	Low	Low
Maglev	High	Medium	High
Jump	Medium	Low	Medium
AnchorHash	High	High	Medium

5   Conclusion and Future Work

In this paper we have studied several hashing algorithms and assessed their applicability to data balancing in distributed databases. For this, we have performed both simulated and real experiments. The real experiments were run using the Unidata platform, an open-source tookit for building MDM solutions. In these experiments we have employed three criteria of applicability, namely uniformity of produced data distribution, the amount of moved records, and computation costs.

Experiments demonstrated that out of seven considered algorithms there are two clear winners—AnchorHash and Maglev. Another two, Linear Hashing and Jump, may have some applicability too.

There are several possible directions for extending this paper. First of all, we noticed some impact of the random number generator on the behavior of some algorithms. In the current paper, we fixed it for all algorithms, but it may be worthwhile to explore this influence. Secondly, it may be interesting to study the impact of varying the parameters of the algorithms. In this paper we have used either the default or the recommended ones, but it is possible that careful tuning may yield positive results.

Acknowledgments. We would like to thank Anna Smirnova for her help with the preparation of the paper.

References

1. Allen, M., Cervo, D.: Multi-Domain Master Data Management: Advanced MDM and Data Governance in Practice, 1st edn. Morgan Kaufmann Publishers Inc., San Francisco, CA, USA (2015)
2. Alon, N., Dietzfelbinger, M., Miltersen, P.B., Petrank, E., Tardos, G.: Linear hashing. Technical report (1997)
3. Chi, L., Zhu, X.: Hashing techniques: a survey and taxonomy. ACM Comput. Surv. **50**(1), 1–36 (2017). https://doi.org/10.1145/3047307
4. DeCandia, G., et al.: Dynamo: amazon's highly available key-value store. SIGOPS Oper. Syst. Rev. **41**(6), 205–220 (2007). https://doi.org/10.1145/1323293.1294281

5. Eisenbud, D.E., et al.: Maglev: a fast and reliable software network load balancer. In: Proceedings of the 13th Usenix Conference on Networked Systems Design and Implementation, pp. 523–535. NSDI'16, USENIX Association, USA (2016)
6. Honicky, R., Miller, E.: A fast algorithm for online placement and reorganization of replicated data. In: Proceedings International Parallel and Distributed Processing Symposium, p. 10 (2003). https://doi.org/10.1109/IPDPS.2003.1213151
7. Honicky, R., Miller, E.: Replication under scalable hashing: a family of algorithms for scalable decentralized data distribution. In: 18th International Parallel and Distributed Processing Symposium, 2004. Proceedings, pp. 96 (2004). https://doi.org/10.1109/IPDPS.2004.1303042
8. Jafarnejad Ghomi, E., Masoud Rahmani, A., Nasih Qader, N.: Load-balancing algorithms in cloud computing: a survey. J. Netw. Comput. Appl. **88**, 50–71 (2017). https://doi.org/10.1016/j.jnca.2017.04.007
9. Karger, D., Lehman, E., Leighton, T., Panigrahy, R., Levine, M., Lewin, D.: Consistent hashing and random trees: distributed caching protocols for relieving hot spots on the world wide web. In: Proceedings of the Twenty-Ninth Annual ACM Symposium on Theory of Computing, pp. 654–663. STOC 1997, Association for Computing Machinery, New York, NY, USA (1997). https://doi.org/10.1145/258533.258660
10. Kuznetsov, S., et al.: Unidata – a modern master data management platform. In: Proceedings of the 1st International Workshop on Data Platform Design, Management, and Optimization (DATAPLAT) co-located with the 25th International Conference on Extending Database Technology and the 25th International Conference on Database Theory (EDBT/ICDT 2022), Edinburgh, UK, March 29, 2022. CEUR Workshop Proceedings, CEUR-WS.org (2022)
11. Lakshman, A., Malik, P.: Cassandra: a decentralized structured storage system. SIGOPS Oper. Syst. Rev. **44**(2), 35–40 (2010). https://doi.org/10.1145/1773912.1773922
12. Lamping, J., Veach, E.: A fast, minimal memory, consistent hash algorithm (2014)
13. Litwin, W., Menon, J., Risch, T.: Lh* schemes with scalable availability (2001)
14. Litwin, W., Neimat, M.A.: High-availability LH* schemes with mirroring. In: Proceedings of the First IFCIS International Conference on Cooperative Information Systems, p. 196. COOPIS 1996, IEEE Computer Society, USA (1996)
15. Litwin, W., Neimat, M.A., Lev, G., Ndiaye, S., Seck, T.: LH*s: a high-availability and high-security scalable distributed data structure. In: Proceedings Seventh International Workshop on Research Issues in Data Engineering. High Performance Database Management for Large-Scale Applications, pp. 141–150 (1997). https://doi.org/10.1109/RIDE.1997.583720
16. Litwin, W., Risch, T.: LH*g: a high-availability scalable distributed data structure by record grouping. IEEE Trans. Knowl. Data Eng. **14**(4), 923–927 (2002). https://doi.org/10.1109/TKDE.2002.1019223
17. Litwin, W., Moussa, R., Schwarz, T.: $LH*_{RS}$-a highly-available scalable distributed data structure. ACM Trans. Database Syst. **30**(3), 769–811 (2005). https://doi.org/10.1145/1093382.1093386
18. Litwin, W., Neimat, M.A., Schneider, D.A.: LH: linear hashing for distributed files. In: Proceedings of the 1993 ACM SIGMOD International Conference on Management of Data, pp. 327–336. SIGMOD 1993, Association for Computing Machinery, New York, NY, USA (1993). https://doi.org/10.1145/170035.170084
19. Loshin, D.: Master Data Management. Morgan Kaufmann Publishers Inc., San Francisco, CA, USA (2009)

20. Mendelson, G., Vargaftik, S., Barabash, K., Lorenz, D.H., Keslassy, I., Orda, A.: Anchorhash: a scalable consistent hash. IEEE/ACM Trans. Netw. **29**(2), 517–528 (2021). https://doi.org/10.1109/TNET.2020.3039547
21. Thaler, D., Ravishankar, C.: A Name-Based Mapping Scheme for Rendezvous. Technical report. https://www.eecs.umich.edu/techreports/cse/96/CSE-TR-316-96.pdf
22. Özsu, M.T., Valduriez, P.: Principles of Distributed Database Systems, 3rd edn. Springer, Cham (2011)

Applications

A New Tool Based on GIS Technology for Massive Public Transport Data

Nieves R. Brisaboa, Guillermo de Bernardo, Pablo Gutiérrez-Asorey[✉],
José R. Paramá, Tirso V. Rodeiro, and Fernando Silva-Coira

Universidade da Coruña, Centro de investigación CITIC, A Coruña, Spain
{nieves.brisaboa,guillermo.debernardo,pablo.gutierrez,jose.parama,
tirso.varela.rodeiro,fernando.silva}@udc.es

Abstract. In this work, we present the design of a novel geographic information tool for the analysis of public transportation data.

The widespread integration of user traveler cards have enabled public transport operators to generate and store large amounts of data related to user movements within the transport network. However, these authorities are seldom equipped to efficiently exploit these data in order to produce a comprehensible analysis of the transport network usage. This is not only due to the sheer amount of data in need of processing, but also because most public transport operators only validate the travel card on boarding, whereas data referring to transfers and alightings are generally unavailable.

Thus, the system we propose not only addresses efficient storage and exploitation of big datasets, but also the reconstruction of complete journeys by using a prediction algorithm to deduce the alighting stop for each boarding. Furthermore, we also provide the transport operators with easy-to-use means of visualizing and analyzing the data through a graphical interface.

Keywords: Geographic information systems · Compact data structures · Big data visualization

1 Introduction and Motivation

The main goal of this work is to build a new tool based on geographic information technology for the analysis of trips in the urban and/or metropolitan public transport network (including metro, commuter train, trams, urban buses, and

This work was supported by the CITIC research center funded by Xunta de Galicia, FEDER Galicia 2014-2020 80%, SXU 20% [ED431G 2019/01 (CSI)]; Partially funded by [RTI-2018-098309-B-C32]; MCIN/ AEI/10.13039/501100011033 [PID2020-114635RB-I00], [PID2019-105221RB-C41], [PDC2021-120917-C21], [PDC2021-121239-C31]; by GAIN/Xunta de Galicia [ED431C 2021/53] GRC; by Xunta de Galicia/Igape [IG240.2020.1.185]; by [IN852D 2021/3] CO3, UE, (FEDER), GAIN, Convocatoria Conecta COVID and by Xunta de Galicia [ED481A/2021-183].

P. Fournier-Viger et al. (Eds.): MEDI 2022, CCIS 1751, pp. 121–135, 2022.
https://doi.org/10.1007/978-3-031-23119-3_9

interurban buses). This application will exploit the fact that currently most trips on public transport begin with the validation of a traveler's card. This validation allows recording a large amount of information about the trips and opens the door to derive the transport needs of citizens, including the global use of the network, the most used stops and their peak hours, the occupation of means of transport, etc. Transport authorities and operators are currently facing major problems in adequately storing all the big data generated (millions of records of data produced daily by traveler cards, generating an extensive history of use), so that later they can be analyzed and exploited efficiently. Besides, due to the particular characteristics of many public transport means, complete information on journeys and transfers is sometimes not available since only entry points to the transport network are stored; e.g., buses and many subways where users do not have to validate their card when alighting.

Despite the existence of algorithms capable of deducing the final stops of the travelers just analyzing their accesses to the transport network [1,2], many transport network administrations are still using on-site surveys to create origin-destination matrices in order to analyze the use of the network made by travelers. These surveys have a high cost, so they are only done sporadically (normally once every several years), and therefore they do not reflect the changing reality of people's mobility. For example, they do not reflect the changes caused by the pandemic context, where the use of public transport has changed radically, they are not even adequate to reflect more gradual changes such as those caused by the appearance of new forms of transport such as electric scooters, rental bicycles, etc.

Thus, although the entities in charge of the transport network potentially have very valuable information on the use of the network, in general, they do not have the appropriate tools for its analysis and exploitation, which limits its usefulness.

This work will tackle three main technological shortfalls in the state of the art:

- Lack of efficiency in the storage, integration and management of travel data in relational databases.
- Difficulties in implementing efficient algorithms to accurately infer final stops of travelers.
- Difficulties in visualizing trajectory data with conventional GIS technology.

2   Related Work

Object trajectories has attracted a lot of attention recently [7] as the number of vehicles and devices equipped with GPS technology or other location means has grown explosively. The collected data have many applications, including traffic management, analysis of human movement, tracking animal behavior, security and surveillance, military logistics and combat, and emergency-response planning [5]. The vast quantities of data, however, can make storing, processing, analyzing them a challenge.

In this work, we present a project that will develop a complete system to collect, store, process, and analyse transportation data. This implies that this system will include a wide variety of technologies and methods from data structures and algorithms [4,12] to sophisticated analysis techniques [1,2,6].

3   Previous Concepts and System Architecture

This section includes some basic concepts. First, we need to precisely define some vocabulary that refers to particular concepts about the normal behavior of a public transport network that will be used in the following sections:

- **Station**: we consider a station as the physical location where one or more lines of any means of transport stop.
- **Line**: this refers to a line providing public transit of any means of transport. It is also important to distinguish between line and **route**. A route is an ordered, consecutive succession of stations that defines a path within a line. At the bare minimum, in the dataset we used, every line has two routes corresponding to the same series of stations in two possible directions.
- **Stop**: with this we are referring to the act of a specific means of transport, following a specific route, stopping at a station and allowing for the boarding and alighting of passengers.
- **Trip and trip-stage**: A trip is the complete journey of a user from one point (called origin) to another (called destination). A trip may require boarding multiple means of transport. A trip-stage starts when the user boards any means of transport at a stop, and ends when the user alights at another stop. Therefore, we define a trip as a sequence of one or more coherent trip-stages. The origin of a trip is the boarding stop of its first trip-stage, and the destination is the alighting stop of its last trip-stage.

Having these concepts and the challenges mentioned above in mind, we combine in the same architecture classic and well-established engineering practices with cutting edge technologies.

The proposed system is a classical full-stack solution composed by three main layers: storage, middle communication and user interface. Figure 1 depicts every module encompassed in its corresponding layer, notice how the modules with the most significant contributions are represented with dashed lines. The following is a brief summary of the goal of each layer:

- **Storage layer.** On the one hand, it is responsible for storing basic data in classical databases (considering the spatial dimensions). On the other hand, this layer is in charge of processing the raw data records, calculating final stops for all the trips and storing the information in enhanced structures designed to reduce space usage while preserving good performance. Next sections will delve into these two advanced components.
- **Communication Layer.** Classic middle tier that serves the information saved in the storage layer to the interface layer.

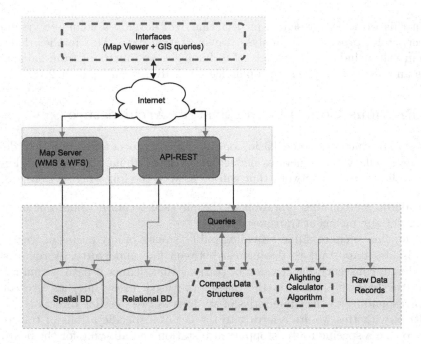

Fig. 1. System architecture with its three distinct parts: storage layer (red), communication layer (green) and user interface layer (blue). (Color figure online)

- **User interface layer.** This layer will provide advanced query interfaces on available public transport travel data. It will provide intuitive interfaces for performing a wide set of predefined but parameterizable advanced queries. For its implementation, we will use standard Web development technologies. Commonly, those results will be presented in the form of maps in the map viewer; in other cases, the results will be presented as lists, statistics, etc. A more detailed description of the contributions on this layer can be found in Sect. 7.

4   Raw Data, Clean Data and Alighting Estimation

All the data used for this project was provided by the Regional Consortium of Transportation of Madrid, Spain.[1] The data encompasses all the transactions of every user card, within any means of public transport, in the city of Madrid, in the years 2019 and 2021.[2] This amounts to a dataset of more than 250 GiB. For reference, just the month of January of 2019 includes 151,094,796 records.

The first major challenge of this work was to interpret, clean and process the raw data. These data consists of twenty-four similarly formatted tables (one for

---

[1] https://www.crtm.es/.
[2] 2020 was excluded from the analysis due to the anomalous use of the network caused by the pandemic scenario.

each month of the years 2019 and 2021) where each record represents one single transaction on the public transport network of Madrid.

A transaction is tied to an user card id, as well as a specific date and time. Transactions also specify the type of user card, the type of user and the type of discount, if any, that was applied to that transaction. Every transaction is also accompanied by a *paypoint code*, that refers to the specific stop where the transaction took place, as well as a validation code that identifies the conditions in which the transaction took place.

The data include all current user cards in the public transport network of the city of Madrid, including user cards for the elderly people, tourist user cards, etc. We also have data regarding single tickets sold not tied to any user card. However, given that the prime interest of this project is to explore the continuous movements of people, we chose to discard such data (corresponding to a 1.3% of the total number of transactions) for the time being.

We were also provided with a set of tables defining the topology of different public transport networks on Madrid, one for each means of transport included in the transactions dataset, those being: subway, suburban train, trolley car, urban bus, and interurban bus. Each record on these tables describes a specific stop in full detail (including name, station and line that it belongs to, exact geographical coordinates, among others) tied also to a *paypoint code*. By using this value, we are able to correlate each transaction with the full information of the stop where it took place.

Figure 2 shows the complete statistics of transactions by means of transport in the month of January 2019. In summary, 46% of the total number of transactions corresponds to subway stations, making that the most used means of transport by a wide margin, with urban bus (24%) being the second most used, followed by suburban train (16%), interurban bus (14%) and, lastly, trolley car (2%).

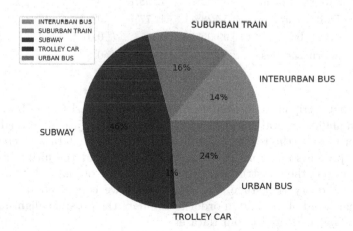

Fig. 2. Transactions per transport means in January 2019, including all types of users.

The distinct validation codes could be used to classify every transaction into three broad categories as follows:

- **Boarding**: the user card was validated when boarding at a specific stop.
- **Alighting**: the user card was validated when alighting at a specific stop.
- **Change of vehicle**: the user card was validated when performing a mandatory change of vehicle in some special stations.

It is important to note that for most means of transport, there is no standardized process for validating the user's card upon alighting, meaning that the vast majority of the available data relates only to boardings without a subsequent alighting. 88% of the total number of transactions are classified as boarding transactions, with 8% being classified as alightings and 4% as changes of vehicle.

Table 1 shows the percentages of every transaction type for every different means of transport. Observe how, for both subway train and trolley car, more than 40% of the transactions are labeled as alightings. This is because for those means of transport users are expected to validate their card upon alighting on most stations. In contrast, on the subway network, only a very small number of stations pertaining to special subway segments that are closed off the regular subway network, demand the users validate their cards on alighting, thus only a 1.64% of the transactions on subway are labeled as such. Finally, as there is never a validation on buses when alighting, there are no transactions labeled as alightings on those means of transport.

Table 1. Percentage of transaction types for each different means of transport

Means of transport	Boardings	Alightings	Changes of train
Subway	90.03%	1.64%	8.33%
Suburban train	57.02%	42.98%	0%
Trolley car	57.23%	42.77%	0%
Urban bus	99.99%	0%	0.01%
Interurban bus	100%	0%	0%

Given that a trip is a concatenation of trip-stages consisting of both a boarding and an alighting, and the state of the data we have just described, it is not possible for most of the real trips to be derived from the data we were provided with. This presents a serious problem for us, given that the ultimate objective of this project is the construction of trips based on the data of boardings and alightings for every trip-stage. For this reason, we concluded it necessary to develop some kind of method in order to estimate the possible alighting stop for every boarding registered in our dataset.

5   Alighting Estimation

In this section, we propose an algorithm based on simple rules with the aim of predicting possible alighting stops for every trip-stage based only on boarding data. Figure 3 shows the general flow of that algorithm.

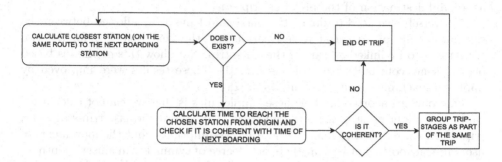

Fig. 3. Flow of the algorithm for the prediction of alighting stop and grouping trip-stages into trips.

In essence, the algorithm consists of two simple steps.

Step 1): For each pair of consecutive boardings, we want to calculate a possible alighting for the first boarding. For this, we search for the closest stop to the second boarding accessible from the first. In other words, it is assumed that there is at least some amount of continuity in the user's movements, so we will start by establishing the possible alighting of a boarding in a stop close to the immediate next detected boarding.

If there is no stop accessible from the first boarding sufficiently close to the second boarding, we determine that it is not possible to deduce an alighting stop at this step, and also that the trip-stage defined by the first boarding and subsequent alighting must constitute an end-of-trip.

Step 2): Assuming Step 1 was successful in identifying a possible alighting stop, it is now necessary to check whether it is reasonable to assume that this created trip-stage can be linked with the next trip-stage as part of a longer trip. In order to accomplish this, we compare the time of the first boarding plus the estimated time needed to arrive at the predicted alighting stop, to the time of the second boarding.

If the two trip-stages are really part of the same trip, the user needs to have been capable of arriving to the estimated destination of the first trip-stage on time to start their second trip-stage without an extended delay, meaning that if there is a difference between the two times larger than a given threshold, it is not reasonable to consider the two trip-stages as part of the same trip, as most likely the user has spent some time on their first destination before starting a new trip.

We apply these rules to every pair of consecutive trip-stages of the same user card. This results in a collection of trips where every trip-stage that is not the last trip-stage of a trip includes a candidate alighting stop.

Of course, it is necessary to define an appropriate threshold for each of these steps. For the first step we need to define the maximum distance beyond which we consider a stop is no longer close enough to another stop for it to be chosen as a valid destination of the previous trip-stage.

Conversely, we need to define the maximum time delay allowed between the estimated time of arrival at a destination and the next boarding for the two trip-stages to be linked as part of the same trip. Note how this method is based on the same continuity principle used in [1]. These results were improved by applying machine learning techniques in [2].

One obvious shortcoming of these simple rules is that we cannot calculate a candidate alighting stop for any trip-stage that is also the last trip-stage of a trip. For this, we need to make yet another assumption about the movements of users in transport, that is, that there is a degree of symmetry in a user's journeys between different days. For example, if a user always starts her day by boarding at a specific stop that we could assume is located near her home, and her last boarding of the day is on an stop that has a possible destination close to that first boarding of the day, it would be reasonable to assume that such destination could be the last alighting stop of the day for this user, and that this last trip of the day represents the act of "going back home".

By searching for regular patterns in the movements of the users and applying this principle of symmetry, it should be possible to further complete the data of the trips.

Furthermore, given that we were provided with information about user's profiles and the types of user cards used for every transaction, we are currently investigating possible alternatives to incorporate this data in the flow of the algorithm in order to enhance the prediction.

To close up this section, note that, while our dataset is certainly lacking in real data about alightings, we can still perform a validation of the prediction algorithm we designed by comparing the alighting stops suggested by the algorithm on trip-stages for which we do have both the boarding and alighting data (this would be true for most-trip stages on suburban train, trolley car, as well as certain segments of subway network).

6   Trip Representation: Aggregated and Disaggregated Data

To face the Big Data 4V, parallelism is the usual choice. However, the basic data structures of most NoSQL systems are not very suitable for the analytical needs of this project, since those systems rely on parallelism with simple data structures like key-value pairs. Instead, our target is to use more complex data structures such as a trajectory data warehouse [9]. Data warehouse is the natural choice for data analysis since they provide grouping capabilities, which includes the possibility of using of pre-computed aggregated data for speeding up queries.

The classical systems based on relational databases or native data warehouse systems need very expensive dedicated hardware to cope with the amount of data that needs to be handled in this project. Therefore, we will use a new alternative based on compact data structures [8], which are capable of storing data in compressed form but, unlike classical compression methods, they allow to extract (decompress) portions of the data without having to decompress the whole dataset. This opens the door to a new computation paradigm, where data reside in main memory all the time in compressed form avoiding disk accesses, thus yielding faster access times [10]. In addition, taking advantage of the savings in space consumption, compact data structures are usually equipped with indexes (also compressed) or precomputed aggregated data, which helps even more to obtain faster query times. In recent years, compact data structures for managing aggregated data have already been designed for multidimensional data [3].

We are already developing a novel data structure for this project inspired by previous compact data structures. One of the most relevant structures in this context is the Topology & Trip-aware Compact Trip Representation (TTCTR) [4], a compact data structure that stores the individual trips of each traveler. Each trip of a traveler is represented as an sorted set of tuples $<s_i, l_i>$, where s_i is the identifier of the stop and l_i the line to which belong. A vocabulary stores the unique identifier that is assigned to each tuple and trips are represented using this vocabulary. TTCTR concatenates and sorts the set of trips and stores it in a modification of the well-known compact and self-indexed data structure Compressed Suffix-Arrays (CSA) [11].

In the experiments, a dataset of 10 millions of trips occupying 165,39 Mb was reduced to 38.16 MB, that is, only 38.16% of the original space.

As this structure works only for individual trips, the authors also proposed T-Matrices (Trip Matrices) [4]. They emerged as an application of image rendering techniques to the particular study of public transport loads. The main idea is to simplify the approach to store aggregated data while speeding up cumulative queries. In the public transport domain, three queries of this type stand out:

- Aggregations by stop (e.g. number of boardings on a specific vehicle at a given stop S).
- Aggregations by time-interval (e.g. number of boardings at any stop of the line on 08/01/2022).
- Aggregations by stop and time-interval (e.g. number of boardings at a given stop S on 08/01/2022).

The simplest way to store the loads of a transport mean would be a classic table, being one axis the actual stops and the other one each journey of a given vehicle, as it is depicted in Fig. 4 (left). However, to solve any of the queries listed above, it would be necessary to iterate through each of the desired cells. That cumbersome process could be avoided by storing the data in an aggregated matrix, summing all values from the top-left corner to the bottom-right corner. Thus, we can compute the total sum of any submatrix in O(1) time

	STOPS								Aggreg.	STOPS							
	0	1	2	3	4	5	6	7		0	1	2	3	4	5	6	7
JOURNEYS 0	0	0	0	0	0	0	0	0	JOURNEYS 0	0	0	0	0	0	0	0	0
1	0	2	1	3	2	1	3	0	1	0	2	3	6	8	9	12	12
2	0	1	1	2	1	0	1	2	2	0	3	5	10	13	14	18	20
3	0	2	2	1	2	0	0	1	3	0	5	9	15	20	21	25	28
4	0	1	3	1	0	3	1	0	4	0	6	13	20	25	29	34	37

Fig. 4. T-matrices. The left table represents the actual gathered values while table on the right represent the same information using an aggregated approach. (Color figure online)

using the formula: $\mathsf{sumSubMat}((x_1, y_1), (x_2, y_2)) \leftarrow M(x_2, y_2) - M(x_2, y_1 - 1) - M(x_1 - 1, y_2) + M(x_1 - 1, y_1 - 1)$.

In the example of Fig. 4, each cell represents the amount of travelers that boarded into a given vehicle during one of its journeys at a given stop. In order to calculate how many travelers boarded on journeys 2 and 3 from stops 2, 3 and 4 (submatrix highlighted in blue) it would be necessary to perform $1 + 2 + 1 + 2 + 1 + 2 = 9$. Yet, using the aggregated solution (right), the same result can be achieved using the formula *sumSubMat* only accessing four cells: $20 - 8 - 5 + 2 = 9$.

One interesting query not tackled in that work is to precalculate the total number of users who used a particular line and their distribution throughout the day. This distribution indicates the percentage of use of a line every hour, in intervals of 15 min or another range of time. For example, a given line X is used on average by 250 people, where 15% boarded between 9:15 and 9:30, 10% boarded between 9:30 and 9:45 and so on. This allows us to know, among other information, the peak hours of the stops and lines, the real use of the lines, population movement patterns, etc.

The structures previously described in this section are not able to resolve this type of queries efficiently. For this reason, we propose a new method based on the creation of a vocabulary that allows us to solve them. The method is based on the fact that the distributions of use will be repeated very frequently between different lines. For instance, many trips in the night hours will have very low or no usage, mornings will have peak usage due to people going to work, etc.

The main idea is to create a vocabulary with the different percentages of distribution that occur in the transportation area during a day. Then, each trip is assigned the number of passengers and also the identifier in the vocabulary of its distribution of use.

7   User Interfaces for Analysis

The systems exploits the trip information using four main user interfaces, that are devoted to different dimensions of the transport management:

- Detailed network usage: this involves analyzing how many passengers have used a given line, or boarded at a given stop, during a specific day or range of days, and how this number of passengers evolves during the day.

– Origin-destination demand: this involves displaying the overall demand by
  users for specific origin-destination trips (i.e. number of full trips, possibly
  including several trip-stages and multiple transport means, that start at a
  given stop and end at a given stop in a given range of dates).
– Anomaly analysis: this involves identifying potential origin-destination pairs
  that have sufficiently high demand but do not have a convenient combination
  of lines to do the trip in reasonable time. This kind of analysis is mainly useful
  for tactical and strategical decisions related to transport planning.
– Accessibility analysis: this involves identifying the connectivity of specific
  stops. The goal of this analysis is to be able to determine, from any given
  stop in the transport network, the average time required to reach any other
  point of the network.

Figure 5 displays a sample interface for detailed network usage analysis. This
is the most detailed query interface, and also answers the most basic usage
analysis queries: The interface displays all the lines for all transportation means,
and users can select a specific line (or, more specifically, a route of that line) in
order to obtained detailed information about the number of passengers using that
specific route. For a given route, user can filter data to a given temporal range,
and to a subset of stops. The system will compute the desired metric for each
selected stop in the line and display the evolution of that metric in predefined
temporal intervals (in the example, 1 h). The metrics available for each stop
include the number of passengers boarding, alighting, switching transportation
means, as well as starting or ending trips in the given stop. This information can
be obtained from aggregated data stored using the data structures introduced
in Sect. 6.

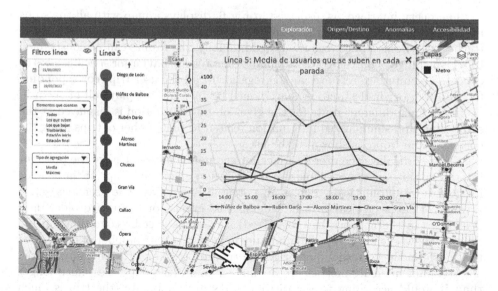

Fig. 5. Network usage analysis interface.

Figure 6 displays the origin-destination analysis interface. This interface provides a higher level analysis of the transportation habits of passengers, as it focuses on characterizing full trips of users. The goal of this interface is to provide an intuitive way to identify the demand on the network between two given points.

Fig. 6. Origin-destination query interface.

The query interface, as displayed in Fig. 6, displays a map with all the stops in the region, where the user may select any two stops as origin and destination, to check the demand between them. Additionally, the results can be constrained to a given range of dates in order to analyze detailed behavior for specific dates or events. The interface retrieves an aggregation of all the user trips corresponding to the given origin-destination, partitioned by line, and displays all of them in the map. Each different route for the given origin-destination is displayed at the same time, and summarized information related to the number of trips using the corresponding route is shown to the user.

Figure 7 displays the user interface for anomaly analysis. Unlike the previous interfaces, this one does not require the selection of specific stops or lines, since its goal is to automatically detect all potential anomalies. The user may select a range of dates (typically, to include specific events, or to discard older data if changes in the network have occurred), as well as specific conditions on the dates (e.g.: weekends only), the minimum demand (i.e. the number of passengers that perform the corresponding origin-destination trip) and the anomaly metric. The relevant anomaly metrics are 3: time anomaly (the trip takes much longer than it would according to the distance), distance anomaly (the trip is much longer than it could be), and vehicle changes (the trip requires changing lines

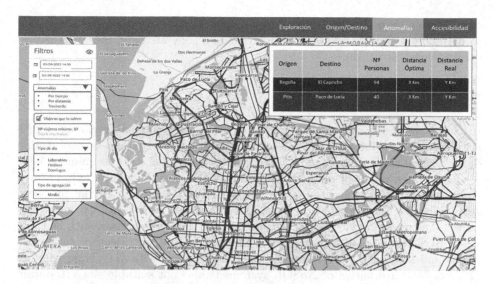

Fig. 7. Anomaly analysis interface.

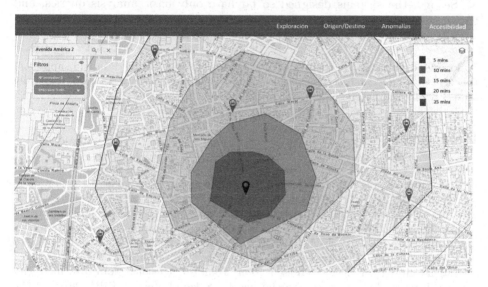

Fig. 8. Accesibility analysis interface.

too many times). Given these parameters, the system computes and displays all the origin-destination pairs that contain an anomaly.

The accessibility analysis interface, displayed in Fig. 8, provides a simple interface to select any stop in the transport network and display the estimated time from the given stop to any other stop in the network. This is displayed in the form of an isochrone map, that isolates the different regions that can be reached in fixed time intervals. Notice that, since different stops may have very different connectivity and transfer times, the interface also provides the ability to manually select the number of isochrone lines generated and the time interval used to construct them.

8   Conclusions

In this work, we have presented an ambitious project under development of a novel geographic information system for the analysis of public transportation data. Several challenges force us to adopt new state of the art techniques.

First, the amount of data being produced at a very high rate daily in large cities and their metropolitan areas. In our case, we include in the proposed system compact data structures to build a completely new trajectory data warehouse. The reason of this choice is that classical data warehouses, such as those based in relational model, first, are not adequate to build a trajectory data warehouse and second, they have serious problems to deal with very large volumes of data. The alternative of NoSQL systems is also not suitable since to obtain aggregated data, all must be done on the fly.

Second, the systems designed so far have only basic analysis queries, and thus, we designed a new set of queries and query interfaces that provide very complex data analysis for a transportation analyst.

Acknowledgments. We wish to acknowledge the cession of the data and support of the *Consorcio Regional de Transportes de Madrid* (https://www.crtm.es/), especially Juan Elices Torrente and Luís Criado Fernández.

References

1. Alsger, A., Assemi, B., Mesbah, M., Ferreira, L.: Validating and improving public transport origin-destination estimation algorithm using smart card fare data. Transp. Res. Part C Emerg. Technol. **68**, 490–506 (2016)
2. Assemi, B., Alsger, A., Moghaddam, M., Hickman, M., Mesbah, M.: Improving alighting stop inference accuracy in the trip chaining method using neural networks. Public Transp. **12**(1), 89–121 (2019). https://doi.org/10.1007/s12469-019-00218-9
3. Brisaboa, N.R., Cerdeira-Pena, A., López-López, N., Navarro, G., Penabad, M.R., Silva-Coira, F.: Efficient representation of multidimensional data over hierarchical domains. In: Inenaga, S., Sadakane, K., Sakai, T. (eds.) SPIRE 2016. LNCS, vol. 9954, pp. 191–203. Springer, Cham (2016). https://doi.org/10.1007/978-3-319-46049-9_19
4. Brisaboa, N.R., Fariña, A., Galaktionov, D., Rodeiro, T.V., Rodríguez, M.A.: Improved structures to solve aggregated queries for trips over public transportation networks. Inf. Sci. **584**, 752–783 (2022)
5. Gudmundsson, J., Laube, P., Wolle, T.: Movement patterns in spatio-temporal data. Encyclopedia of GIS **726**, 732 (2008)
6. Kopczewska, K.: Spatial machine learning: new opportunities for regional science. Ann. Reg. Sci. **68**, 713–755 (2021). https://doi.org/10.1007/s00168-021-01101-x

7. Mahmood, A.R., Punni, S., Aref, W.G.: Spatio-temporal access methods: a survey (2010–2017). GeoInformatica **23**(1), 1–36 (2018). https://doi.org/10.1007/s10707-018-0329-2
8. Navarro, G.: Compact Data Structures: A Practical Approach. Cambridge University Press, USA (2016)
9. Pelekis, N., et al.: Towards trajectory data warehouses. In: Giannotti, F., Pedreschi, D. (eds) Mobility, Data Mining and Privacy. Springer, Heidelberg (2008). https://doi.org/10.1007/978-3-540-75177-9_8
10. Plattner, H., Zeier, A.: In-Memory Data Management: Technology and Applications. Springer, Heidelberg (2012). https://doi.org/10.1007/978-3-642-29575-1
11. Sadakane, K.: New text indexing functionalities of the compressed suffix arrays. J. Algorithms **48**(2), 294–313 (2003)
12. Zheng, Y., Zhou, X. (eds.): Computing with Spatial Trajectories. Springer, New York (2011)

Mobile Application Code Generation Approaches: A Survey

Shaymaa Sayed El-Kaliouby[1,2]([✉]), Ahmed H. Yousef[1,3], and Sahar Selim[1,2]

[1] School of Information Technology and Computer Science, Nile University, Giza, Egypt
{Selkaliouby,Ahassan,SSelim}@nu.edu.eg
[2] Center for Informatics Science, Nile University, Giza, Egypt
[3] Faculty of Engineering, Ain Shams University, Cairo, Egypt

Abstract. With the extensive usage of mobile applications in daily life, it has become crucial for the companies of software to develop applications for the most popular platforms such as Android and iOS in the shortest possible time and at the lowest possible cost. However, ensuring consistent UIs and functionalities among cross-platform versions can be challenging and costly since different platforms have their own UI controls and programming languages. Also, when cross-platform tools are used, it is always time consuming to learn a new language. Many solutions were proposed to achieve the native performance. In this paper, two categories are surveyed: Code generation and code mapping which is also known as code-to-code conversion. The code generation approach is concerned with generating user interfaces for applications utilizing deep learning and image processing techniques. Meanwhile, the code-to-code mapping method maps native code to the appropriate platform. In addition, proposals for further investigations are made.

Keywords: Cross-platform · User Interface (UI) · Code-to-code conversion · Code generation · Mapping · User Experience (UX) · Graphical User Interface (GUI)

1 Introduction

Today, developing a mobile application typically involves creating two versions, one for Android and one for iOS. To attract clients, it is critical to have a user-friendly and consistent user interface across both platforms (Android & iOS) as well as the web version of the application. Also, it is very important to have the same functionality on all platforms. To address this challenge, Cross-platform development was proposed, as the applications run on multiple platforms when developed once. That means Cross platform development is based on creating software applications that can be deployed on more than one platform. Examples of the tools that are based on cross platform development are the following: Xamarin [1], Cordova [2], React Native [3], Flutter and Ionic. Unfortunately, these tools still can't reach the performance of the native applications.

Cross platform mobile development was the subject of many surveys and research papers. Each one of these papers have categorized the approaches based on some studies

P. Fournier-Viger et al. (Eds.): MEDI 2022, CCIS 1751, pp. 136–148, 2022.
https://doi.org/10.1007/978-3-031-23119-3_10

and analysis. The proposed solutions with approaches to generate the user interface code or user interface code integrated with the backend code or map the existing code of specific platform to another platform. All these solutions are presented and categorized in this survey. The classification is carried out to select the best methodology for generating cross-platform applications which will be developed once and run-on multiple platforms but with the performance of natively developed applications which is higher than the cross-platform developed applications.

The contribution of this study can be listed as follows:

• It categorizes the approaches used to generate code from user interface screenshots/user interface sketching. Also, the approaches that are used to convert native code to another native code or generate code from user interface screenshots/user interface sketching.
• It suggests the best approach and methods to be followed in order to achieve the optimal outcome for simpler development and better performance.

The rest of this paper is organized as follows: Sect. 2 presents the related work. Then the first approach which is GUI automated code generation is presented in Sect. 3. Section 4 presents the code mapping (code-to-code conversion). Code-to code conversion approach is categorized into two sections.

The first section is code conversion without the UI and the second section is backend code conversion integrated with UI or UI code mapping. Section 5 discusses the two approaches and provides the pros and cons of each approach. Finally, Sect. 6 concludes the survey.

2 Related Work

Numerous researchers have published many articles about mobile development and the techniques employed in it. The main reason behind the research is to have the high performance of the native applications and reduce development time without having to learn new languages to develop the applications for all platforms. Some developers also decided to choose the method of converting from native-to-native code or generating native code from sketches or UI screenshots. This survey paper focuses on demonstrating the approaches to generate native code from mobile applications user interface screenshots or user interface sketching and the approaches used to translate code from one programming language to another.

In [4], the authors presented a comparison of the approaches for cross platform mobile development based on different criteria. The approaches were categorized as follows:

1. *Web approach*: it is an approach for mobile devices based on web browsers. The applications are created with the help of HTML, CSS, JavaScript, etc.
2. *Hybrid approach*: this approach is a combination of using web technologies and native technologies. Examples of frameworks that use this approach are Ionic, Angular... etc.

3. *Interpreted approach*: it is the approach where the developers use common language like JavaScript to write the code of user interface and generate the equivalent for native component for each platform. The native features are provided by an abstract layer that interprets the code on runtime across different platforms to access the native APIs.
4. *Cross-compiled approach or generated approaches*: they are the approaches where the developers write codes with the use of any common programming language. Then these codes are transformed by cross compilers to a specific native code. Example of this approach is android, iOS and Xamarin.
5. *Model-driven approach*: Based on Model Driven Architecture, this method enables developers to describe applications at a high level without having to deal with the details of the low level. The model driven approach has the three following models:

   5.1 A *software model* that is independent of the particular platform being used to implement it is known as a platform independent model.
   5.2 A *platform specific model* is the one that includes the specifics of how the system utilizes a specific kind of platform.
   5.3 *Platform independent model (PIM) to platform-specific model (PSM)* conversion which is known as model transformation.

They have observed that, in comparison to the other ways, the model-driven approach has the most potential for expansion more than the other approaches.

Another survey was published where the authors were comparing and evaluating the cross-platform mobile development tools [5]. This research outlines various factors beyond portability considerations that should be considered when selecting a suitable cross-platform tool for development. The following design factors were examined between the three techniques by the authors:

1. Native Approach.
2. Mobile Web Approach.
3. Cross-platform Approach.

The decision criterion is based on the following:

1. Quality of UX.
2. Quality of the Apps.
3. Intended Audience/Users.
4. Cost of Developing the application.
5. The application security.
6. Supportability, Maintainability, time to market and extending the application after development.

They discovered that the native approach has superior UX quality compared to the mobile web approach and the cross-platform approach. Along with application quality and security, the native method approach is higher than the web and cross-platform. The cost of native development is the one of the issues they found, as it is more expensive than mobile web approach and cross-platform approach. It also limits the number of users because native development is platform-specific.

In [6], the authors made a comparison between the native and the cross-platform frameworks. They tested the start-up-time, application size, memory usage and CPU usage. They found that the applications deployed using native development are consuming less memory and CPU usage. Also, the startup time was faster in native than the cross platform developed applications.

This study presents the approaches of generating the native mobile applications code. It differs from the other studies as it doesn't discuss the difference between native and cross-platform development because already natively developed applications are faster than cross-platform developed applications. So, we present the different approaches for generating native applications and present the optimal technique or method to generate applications with better performance and lower costs. The paper also addresses UI conversion, which is quite uncommon in other studies.

3 Automated Code Generation

This section illustrates the approaches and methods used to automatically generate a mobile application native code or web application code using neural networks and deep learning techniques. The native code is generated by just providing a user interface screenshots or sketching.

3.1 GUI Automated Code Generation

In [7], Chen et al. created a framework that collects user interface screenshots from Android/iOS pages and generates GUI code for the target platform (Android/iOS). Figure 1 shows the input, output of the generator and the process.

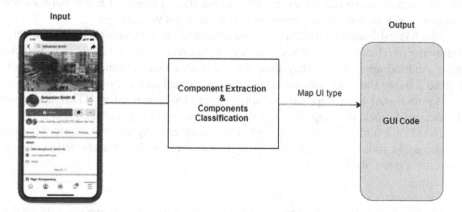

Fig. 1. Automated GUI code generation process

The authors used two image processing techniques, Canny edge detection algorithm and wide range detection algorithm. Canny edge detection is a method for extracting meaningful structural information from various vision objects and lowering the amount of data required for processing. The two techniques are used to capture most of the image's edges. Additionally, edge dilation can be used to merge adjacent pieces and enlarge the boundaries, eliminating any gaps that may exist. Then they used Convolutional Neural Network (CNN), they applied this deep neural network to analyze the image and do image classification. The following steps illustrates how the code is generated.

1. *Component identification*: Which is extracting the components using image processing. Then identify the corresponding GUI code using deep learning.
2. *Component type mapping*: This is mapping each component type with the corresponding platform (Text in android to Text in iOS).
3. *GUI Code Generation*: after mapping, the GUI code for the needed language is being generated. In their GUI code generation, they use CNN classification (Convolutional Neural Networks) to classify the components that need to be mapped.

They used testing tools UIAUTOMATOR and STOAT for Android apps and ID EVICEINSTALLER for iOS just to extract the components of the application screens.

In [8], the authors have developed STORYDROID. STORYDROID is a system or tool that produces storyboard of Android apps by doing the following:

1. *Transition extraction*: obtaining an activity transition graph.
2. *Rendering the UI pages*: in UI rendering, it takes all the layouts (dynamic, hybrid or static) and convert it to static layout to render the pages that the users are interacting with.
3. *Semantic name*: inferring names for obfuscated activities.

In [9], Mohian et al. provided native app code by freehand UI sketching called Doodle2App. Their paper leverages the recent Google Quick draw, which is a dataset of 50M sketch stroke sequences, in order to pre-train a recurrent neural network and retrains it with sketch stroke sequences they collected it via Amazon.

Mechanical Turk. Doodle2App is a website offers a paper substitute. It means that a drawing interface with an interactive UI preview and can convert sketches to a single-page Android application. They used the method of convolutional Neural Network. Figure 2 shows the input and output of the implemented methodology.

Agrawal et al. have proposed a solution where a tool is generating HTML code from the designed mockups using deep learning and machine learning techniques [10].

In [11], the authors presented a novel method for creating automatic cross-platform mobile applications utilizing image processing techniques. The UI generation process consists of three phases:

1. *Image analysis*:
    The images that will be processed via image processing are analyzed during this phase. The procedures used to create an application's source code. There are three main steps in the analysis:

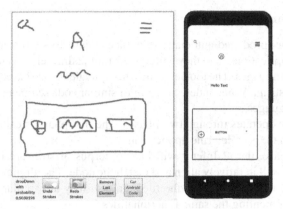

Fig. 2. Sketch and generated android app [11]

  1.1 Input image processing
  1.2 Input image validation
  1.3 Elements identification

2. *Configuration:*
   The major goals of this phase are to configure the application and choose the required platform in order to produce the native source code. The following steps are involved at this phase:

  2.1 Platforms selection
  2.2 Application configuration
  2.3 XML-based document generator

3. *Source Code Generation:*
   ImagIngDev generates the source code for the application at this stage. The processes or the steps in this process change the data that was received throughout the analysis and configuration phases.

  3.1 Process XML-based document.
  3.2 Project files generator and packager.

4 Code to Code Conversion

This section shows the used approaches to convert from code to another code or mapping the code. It has two categories, first one which is mapping just the backend code of the application without the interface, for example converting from swift to kotlin. Second category is mapping the source code to the destination including the backend and the user interface.

4.1 Code Mapping Without UI

In [12], the authors produced automatic inference of java-to-swift translation rules for porting mobile applications, and they called their tool j2sInferer. Given a software code in java and swift languages. The tool identifies the equivalent code based on the similarity of the braces and string. When it detects a pair of similar code segments it creates syntax trees for both languages.

The j2sInferer operates through two phases. The first phase is called rule inference and the second phase is called rule application.

In the first phase, the j2sInferer will have a corpus of software in java and swift languages, then it will iteratively align and match the code. The alignment and matching of the code use to be done by using the file names, if they are similar, the tool assume that they are implementing the same functionalities.

In the second phase, the tool will have a java program to generate a translated swift version.

So, it performs something called statement-to-statement translation.

In [13], Vendramini et al. used trans-pilers (a Tool that converts from language to another language). They worked on converting from swift to kotlin. Their proposed trans-piler is characterized by the following:

1. *Prioritizes its output code's maintainability*: This means the generating the output should be easy to understand, in which the developer that understand the input code, also can understand the output code.
2. *Depends only on maintainable or removable libraries*: This means that the output code must depend only on the small libraries to have easy of maintenance or removal by the developers.
3. *Ensures its output code offers idiomatic APIs*: generated APIs should follow common platform and language conventions and be directly accessible from platform-specific code, without requiring interoperability layers.
4. *Ensures its output code can directly access platform-specific code*: platform-specific code written in both the input and output languages must be accessed directly from the shared code.

In [14], Ahmed et al. provided a tool that converts from swift to java using trans-compilation approach. The tool is based on ANTLR (Another Tool for Language Recognition). The process of conversion was as follows:

1. Select swift as the input language when configuring the ANTLR tool in the java IDE (IntelliJ is used in this research).
2. It is important to inspect the parse tree for each sentence in order to be able to alter the parser code, therefore test the grammar rules generated by ANTLR with some swift code statements.
3. In order to develop java code that complies with the syntax rules for each type of statement, override methods in the ANTLR parser code.
4. Then finally, overridden methods act as translator code, producing converted java code from swift code as an input.

When they tested the solution across six different applications, the accuracy was on average 91.7%.

4.2 Code Mapping with the UI

In [15], Lamhaddab et al. proposed a solution of porting the mobile applications from iOS to Android using model-driven reverse-engineering (MDRE) approach. The implemented tool called iSpecSnapshot. It also houses the mapping between the SDKs APIs of the two platforms. So, the tool is providing the following services:

1. Parsers will allow the generation of platform specific models from iOS src code (PSM models from the iOS mobile app source code). The PSM models correspond to Xcode project structure, programming language (C, C++, Objective-C, and Swift), xib file, plist file, strings file, etc.
2. Extractors examine the PSM models generated by the parsers. Name, icon, build, device orientation, connected frameworks, and screen workflow or UI components are all used by the PIM model to identify specific information on the iOS app.
3. Transformers analyze the Platform Independent Model and initiate a cyclic transformation process to create the documentation Platform Specific Model and the Android PSM model. The PSM model provided by SFD depicts the entire structure and content of the specific requirements documentation, including the cover page, summary page, screen page, and framework dependencies and mapping. The structure of the intended Android UI project is depicted by the Android app platform specific model.
4. Generators shift the generated models around before generating the Android UI skeleton.

In [16], UITrainDroid tool have been implemented to transform the GUI of the projects from iOS 10 to Android 7.0. The authors are following some steps to mapping process. First step is generating layout tree for each page or screen of the source application. In this step, there are prepared scripts for the two versions of the application (Android & iOS), these scripts are designed to execute functions in the same order in order to obtain similar pages on the Android and iOS. Then the second step is module construction. Module construction divides the layout trees of the Android and iOS into modules. The authors in this process are just making sure that all the iOS controls can be mapped to android control with all the modules.

Third step is mapping iOS control to Android controls.

In order to handle the positioning of the controls, a control's position is used. A control position is a two-tuple key-values (x, y) showing how the control is away from the left and top border. For example, if a control position is presented as (0.2, 0.4) it means that the control is 0.2 away from the left border and 0.4 away from the top border.

The third step of mapping is mapping the iOS and Android modules. Since one iOS module can be mapped to many android modules, the authors have implemented a flag field for each pai of modules to record the time of its occurrence.

Storyboard file transformation is mapping the relation of modules in the database and transform storyboard and xib files (iOS page codes) to GUI code of android. The attributes are classified into static attributes and dynamic attributes. First, storyboard/xib files are collected from the source program and encoded to the page layout tree. Then a search process is being applied in the database. The search process is for the iOS module of the stored mapping relations. So, the modules are similar if the two modules have the same structure, and the controls have the same position. Then the last step in the transformation is the construction. The android layout codes are constructed by combining the generated android modules with the obtained information of storyboard and xib files.

El-Kassas et al. have produced a solution using Model Driven Development (MDD) and Cross-Compilation [17]. This solution is implemented to convert from Windows phone to Android platform.

The solution architecture consists of the three blocks:

1. *Abstract Model*: This model is representing the mobile application. Also, this model has the documents of XML in order to handle the UI conversion with the resource files.
2. *AppModeler*: This module takes the whole project which is the whole source code of windows phone 8 app and transform it to XML and resource file (abstract model). This module is implemented separately for each supported source platform.
3. *AppProducer*: Finally, this module transforms the abstract model project to the corresponding project. This module is implemented separately for each supported platform using component-based approach.

This solution is enhanced in [18], where the authors implemented the AppModeler and AppProducer separate from each other so if there is any update needed in any of the platforms, it can be performed without affecting the other platform. Then El-Kassas et al. extended the solution by adding more configuration files like the database and XSLT. Also, the tool is enhanced where it can generate full applications ready to be published.

In [19], Hamza et al. proposed a conversion tool named TCAiOSC. The tool converts Android application to iOS applications using trans-compiler-based approach.

The tool supports code and GUI code transformation. Means that it takes a whole android application that contains XML and java files and convert it to iOS application that contains swift and storyboard files.

So, the conversion process goes as the following:

- The project files are categorized according to the controller. The main unit is then given the java and XML files individually.
- Receiving the java files by the code conversion unit from the main unit is the second stage. The code's parse tree is created by the lexer and parser and sent to the converter.
- The Code Converter leverages the resources in the database as it moves through the parse tree node by node. This results in a Swift file for every java file and a viewController for every Activity.
- Then the Code Conversion Unit sends the Swift files and some data about the Activity files to the Main Unit, such as the Layout file name corresponding to that Activity file and the names of Activity objects with their respective Layout IDs. The UI Conversion Unit will utilise this information to connect the viewController files to the Storyboard.

All Scenes are combined into a single Storyboard file and sent to the Main Unit.

Finally, the Swift and Storyboard files, as well as the new manifest and resources files, are assembled in the iOS project.

In [20], El-Kaliouby et al. enhanced the tool produced by Ahmad in [14]. In this research, the authors focused on adding the features of converting the user interface components from Swift to java. Means that the approach is supporting the conversion of the user interface component of the backend, and it follows the same process as the swift to java converter. When they compared between their solution and the old solution, they found that the accuracy of the UI-unsupported solution was on average 71.8%, whereas the accuracy of the new solution was 88.9% when tested on five applications.

Table 1 summarizes the approaches used in studies in literature for both categorizations, the automated code conversion and code mapping.

Table 1. Summary of code generation approaches

Category	Author	Approach	Source	Destination
Automated code generation	Sen Chen et al. [7]	Image processing & deep learning	UI **Screenshots** of iOS/Android	UI **Pages** of Android/iOS Apps
	Sen Chen et al. [8]	Transition extraction, UI page rendering, Semantic name inferring	Android app	Storyboard
	Mohian et al. [9]	Convolutional Neural Network	UI Sketch	Android/IOS Apps
	Agrawal et al. [10]	Deep & Machine Learning	UI Mockups	HTML Code
	ROSALES-MORALES [11]	Image processing	UI Images	XML document
Code mapping	El-Kassas et al. [17, 18]	Trans-compilation	Windows phone	Android
	Ahmad et al. [14]	Trans-compilation	Swift	Java
	Khalid Lamhaddab et al. [15]	Model-driven reverse-engineering	IOS App	Android APP
	Vendramini et al. [13]	Transpiler	Swift	Kotlin
	Ji et al. [16]	Mapping	IOS 10	Android
	Hamza et al. [19]	Trans-compilation approach	Android	iOS
	El-Kaliouby et al. [20]	Trans-compilation approach	SwiftUI	Java (UI for android)

5 Discussions and Recommendations

Cross-platform development solutions are recommended when an application is devoted to more than one platform, but at the same time, learning new languages and losing the performance of native development is a crucial issue. Therefore, conversion from native-to-native platform or generating a code for android or iOS was a solution to reach the native performance and at the same time avoid learning a new language to implement. This study presented two approaches used in transforming the code and generating the code to produce mobile applications. Table 2 presents the pros and cons of the two approaches.

As aforementioned, the code-to-code conversion outperforms code generation in many aspects. The developer can easily deal with the backend code not like the auto-mated code generation where they can just deal with a sketch or UI screenshots without accessing any functionality from the backend, and it will be so difficult to handle the transition in all pages and the actions of anything in the application. Therefore, this study recommends the use of code-to-code conversion or code mapping. In addition, it is recommended that the code conversion solutions be improved to address the issue of converting certain functionality, such as the UI conversion in the front-end section and the mapping of web-services and libraries in the back-end section in order to have a full functionality generated applications.

Table 2. Comparison between automated code generation and code-to-code conversion

	Automated code generation	Code-to-Code conversion
Pros	• Ease of visualization • Ease of generating a code from basic stuff (Sketch or screenshots or mockups). So, no need to have code to be mapped, just screenshots or sketching	• Understanding the functional behavior of both languages and platforms • Ease of extracting the code of transition between pages • Easier to map the components from language to another
Cons	• Transition extraction: cannot handle the code of transition which means that it cannot extract it • Cannot handle dynamic layouts: the layouts must be converted to static so it can convert it • There is no such a technique that maps and transfers the GUI • Cannot extract accurate coordinate position	• Difficulty of Automating the conversion of some functionalities in the user interface because of the different architecture between the two platforms (Android & iOS) • Difficulty of encoding swift or android files. So, it needs more advanced encoding scheme and related to compiler • Some conversion tools are just converting from windows phone to android. We are not sure that windows phone platform is still used by any mobile nowadays as even Nokia phone now is using android OS • Some critical features are missing in the presented solutions, such as the conversion of web services

6   Conclusion

This study produced the mobile applications code generation approaches. It is categorized in this survey to two categories. First category is the code generation which is generating the code from sketches of user interface screenshots using CNN and image processing techniques. Second category is the conversion of specific code to another code. Our findings illustrated that the code-to-code conversion is better methodology than the automated generation due to the limitations mentioned in the previous section.

To the best of our knowledge, this is the first study that introduced the researches that tackle the problem of the user interface code generation and code conversion for mobile applications. Also, this type of categorization mentioned in Sect. 3 and 4 is done for the first time in this study. It also discussed the problems with each category and highlighted the areas that need more study (Mobile application User Interface code generation/conversion). In conclusion, this study aims to find the optimal method for developing cross-platform applications that can operate on various platforms without sacrificing the native performance.

References

1. Xamarin: Open-source mobile app platform for .NET. https://dotnet.microsoft.com/en-us/apps/xamarin. Accessed 14 Jun 2022
2. Apache Cordova: https://cordova.apache.org/. Accessed 04 Jun 2022
3. Biørn-Hansen, A., Rieger, C., Grønli, T.-M., Majchrzak, T.A., Ghinea, G.: An empirical investigation of performance overhead in cross-platform mobile development frameworks. Empir. Softw. Eng. **25**(4), 2997–3040 (2020). https://doi.org/10.1007/s10664-020-09827-6
4. Latif, M., Lakhrissi, Y., Es-Sbai, N.: Cross platform approach for mobile application development: a survey. In: 2016 International Conference on Information Technology for Organizations Development (IT4OD 2016), pp. 1–5 (2016). https://doi.org/10.1109/IT4OD.2016.7479278
5. Dalmasso, I., Datta, S.K., Bonnet, C., Nikaein, N.: Survey, comparison and evaluation of cross platform mobile application development tools. In: 2013 9th International Wireless Communications and Mobile Computing Conference (IWCMC 2013), pp. 323–328 (2013). https://doi.org/10.1109/IWCMC.2013.6583580
6. Nawrocki, P., Wrona, K., Marczak, M., Sniezynski, B.: A comparison of native and cross-platform frameworks for mobile applications. Computer **54**(3), 18–27 (2021). https://doi.org/10.1109/MC.2020.2983893
7. Chen, S., Fan, L., Su, T., Ma, L., Liu, Y., Xu, L.: Automated cross-platform GUI code generation for mobile apps. In: 2019 IEEE 1st International Workshop on Artificial Intelligence for Mobile (AI4Mobile), pp. 13–16. IEEE (2019). https://doi.org/10.1109/AI4Mobile.2019.8672718
8. Chen, S., et al.: StoryDroid : automated generation of storyboard for android apps. In: 2019 IEEE/ACM 41st International Conference on Software Engineering (ICSE), pp. 596–607 (2019). https://doi.org/10.1109/ICSE.2019.00070
9. Csallner, C.: Doodle2App: native app code by freehand UI sketching. In: Proceedings of the IEEE/ACM 7th International Conference on Mobile Software Engineering and Systems (MOBILESoft 2020). Association for Computing Machinery, New York, pp. 81–84 (2020). https://doi.org/10.1145/3387905.3388607

10. Agrawal, S.A., Suryawanshi, S., Arsude, V., Maid, N.: Artificial intelligence based automated HTML code generation tool using design mockups. J. Interdiscip. Cycle Res. **12**(III) (2020)
11. Rosales-Morales, V.Y., Sánchez-Morales, L.N., Alor-Hernández, G., Garcia-Alcaraz, J.L., Sánchez-Cervantes, J.L., Rodriguez-Mazahua, L.: ImagIngDev: a new approach for developing automatic cross-platform mobile applications using image processing techniques. Comput. J. **63**(5), 732–757 (2020). https://doi.org/10.1093/comjnl/bxz029
12. An, K., Meng, N., Tilevich, E.: Automatic inference of java-to-swift translation rules for porting mobile applications. In: Proceedings of the 5th International Conference on Mobile Software Engineering and Systems, pp. 180–190 (2018). https://doi.org/10.1145/3197231. 3197240
13. Vendramini, V.J., Goldman, A., Mounié, G.: Improving mobile app development using transpilers with maintainable outputs. In: Proceedings of the 34th Brazilian Symposium on Software Engineering, pp. 354–363 (2020). https://doi.org/10.1145/3422392.3422426
14. Muhammad, A.A., Mahmoud, A.T., Elkalyouby, S.S., Hamza, R.B., Yousef, A.H.: Transcompiler based mobile applications code converter: swift to java. In: 2020 2nd Novel Intelligent and Leading Emerging Sciences Conference (NILES), pp. 247–252. IEEE (2020). https://doi.org/10.1109/NILES50944.2020.9257928
15. Lamhaddab, K., Lachgar, M., Elbaamrani, K.: Porting mobile apps from iOS to android: a practical experience. Mob. Inf. Syst. (2019). https://doi.org/10.1155/2019/4324871
16. Ji, R., Pei, J., Yang, W., Zhai, J., Pan, M., Zhang, T.: Extracting mapping relations for mobile user interface transformation. In: Proceedings of the 11th Asia-Pacific Symposium on Internetware, pp. 1–10 (2019). https://doi.org/10.1145/3361242.3361250
17. El-Kassas, W.S., Abdullah, B.A., Yousef, A.H., Wahba, A.: ICPMD: integrated cross-platform mobile development solution. In: 2014 9th International Conference on Computer Engineering & Systems (ICCES), pp. 307–317. IEEE (2014). https://doi.org/10.1109/ICCES.2014. 7030977
18. El-Kassas, W.S., Abdullah, B.A., Yousef, A.H., Wahba, A.M.: Enhanced code conversion approach for the integrated cross-platform mobile development (ICPMD). IEEE Trans. Softw. Eng. **42**(11), 1036–1053 (2016). https://doi.org/10.1109/TSE.2016.2543223
19. Hamza, R.B., Salama, D.I., Kamel, M.I., Yousef, A.H.: TCAIOSC: application code conversion. In: 2019 Novel Intelligent and Leading Emerging Sciences Conference (NILES), vol. 1, pp. 230–234. IEEE (2019). https://doi.org/10.1109/NILES.2019.8909207
20. El-Kaliouby, S.S., Selim, S., Yousef, A.H.: Native mobile applications UI code conversion. In: 2021 16th International Conference on Computer Engineering and Systems (ICCES), pp. 1–5. IEEE (2021). https://doi.org/10.1109/ICCES54031.2021.9686093

TOP-Key Influential Nodes for Opinion Leaders Identification in Travel Recommender Systems

Nassira Chekkai$^{(\boxtimes)}$ and Hamamache Kheddouci

LIRIS Laboratory, Lyon 1 University, Lyon, France
{nassira.chekkai,hamamache.kheddouci}@univ-lyon1.fr

Abstract. Travel recommender systems, also called (TRS) have recently gained significant attention in the research and industrial communities. These systems aim at identifying the travellers preferences and providing adequate suggestions to them whenever and wherever they want. Thus, TRS are very helpful for travelers, particularly, when they visit a place they have never been to before. Opinion Leaders based-technique attempts to identify the set of most important delegates who can represent as many TRS users as possible to alleviate the cold start user problem when a new user is registered to the system and has no ratings yet and the cold start item problem when a new item is added to the system and has no interactions yet. In this paper, we propose two graph based approaches for Opinion Leaders Detection in Travel Recommender Systems. The first is based on the Minimum Cover Vertex identification, while the second uses the Fragmentation method to detect the set of most influential nodes in the recommender system graph. The obtained experimental results confirm the effectiveness of our proposal.

Keywords: Travel recommender systems · Opinion leaders · Context awareness · Cold start problem · Graph theory · Minimum cover vertex · Fragmentation · Cut nodes · Similarity

1 Introduction

Nowadays, mobile devices are becoming our daily companions. We use them to achieve appropriate services and relevant information at anytime and anywhere. Moreover, the rapid growth of mobile applications has provided a huge collection of information of different types (texts, images, sounds, videos, etc.).Thereby, mobile recommender systems have emerged to recommend the right service or information to the right mobile users whenever and wherever they want; by predicting the "rating" a user would give to an item he had not yet considered [1, 2].

Recommender systems generate recommendations in one of two ways, namely collaborative or content-based filtering. Content based-technique generates recommendations by comparing the content of items to the user profile [3], while Collaborative Filtering focuses on the analysis of users' behaviors and reviews to suggest to them a set of interesting items that have been well rated by users having similar preferences and interest centers. Collaborative filtering is considered the most successful and widely used recommendation technique [2, 4].

© The Author(s), under exclusive license to Springer Nature Switzerland AG 2022
P. Fournier-Viger et al. (Eds.): MEDI 2022, CCIS 1751, pp. 149–161, 2022.
https://doi.org/10.1007/978-3-031-23119-3_11

Furthermore, since a majority of the adult population spends actually a great deal of time outside home and like to travel, Travel Recommender Systems (TRS) have recently attracted a great deal of attention of researchers and automotive industries. Indeed, TRS are mobile recommender systems that aim at suggesting relevant items to travelers and tourists. These items represent their Point-Of-Interests (POIs); such as restaurants, hotels, cafeterias, etc.

In addition, Travel Collaborative Recommender Systems (TCRS) occupy a major place in the fields of Information Filtering and Social Networking, and are largely invested in modern information systems. In fact, TCRS have a special recommendation policy as they recommend to a traveler items may be of his interests based on the ratings left by other travelers having similar tastes. These travelers are called Acquaintances. The ultimate goal is to increase the tourist satisfaction by minimizing the search time while suggesting appropriate POIs.

Indeed, TRS are used to provide real time suggestions to tourists and travelers. The generated recommendation depends on the user preference and his context aware information (current situation and time) which change along the time (e.g. the choice of restaurants depends on the relevance of restaurants, the time, the weather and the location of the traveler as the TRS should recommend to him the closest ones). Thus, the system needs to collect and analyze information about the environment and items.

One crucial issue in TCRS is the Cold Start Problem which includes two key aspects: new user and new item. Cold Driver is a new comer who enters the system and cannot get appropriate items. Indeed, the system is unable to predict recommendations for this driver since it has little knowledge about him [5]. In addition, Cold Item is a new item that cannot be recommended since it has no ratings yet [1].

To alleviate cold start problems in recommender systems, most studies present Hybrid recommender systems that combine content based and collaborative filters to overcome their disadvantages [6, 7]; moreover, Ontology-based filtering incorporates the semantic knowledge with the user preferences [8], Fuzzy logic-based filtering uses association rules in order to predict more relevant items [9], Clustering based filtering identifies clusters of similar drivers and/or items using the similarity measures [10], and decision tree based filtering takes into account of the tree structure to effectively generate recommendations [11].

In recent years, considerable research studies in the field of recommender systems have focused on how a set of key users may influence other users by their positive or negative ratings. These leaders can also help studying the future impact of adding new item in the system.

This paper presents two approaches for discovering the most important opinion leaders in TRS, by selecting the set of influential travellers who would help new tourists to get relevant recommendations. For this, we use graph parameters and context information. We consider key vertices, ratings and context aware conditions of users to select the most influential opinion leaders.

The rest of this paper is structured as follows: Sect. 2 reviews related works. Section 3 explains our motivation. Section 4 presents and details our contribution. Section 5 validates our proposal by presenting and discussing the obtained results. Conclusion and perspectives are given in Sect. 6.

2 Related Work

Opinion leaders are users having high assessments influence within their social communities. They influence their friends' circles by their positive or negative opinions on the recommender system items. Indeed, the users generally prefer to visit POIs that were highly rated by some key users or friends rather than travelling to other places.

The identification of opinion leaders in recommender systems has been studied in many research studies. [12] focused on inferred models of influence and the number of rated items to measure and evaluate the stability of recommender systems. [13] treated the concept of "influence" in collaborative filtering using an influence discrimination model. The latter approach detected the set of key users by studying the effect of their removal from the recommender system. [14] employed data mining techniques; namely linear regression and decision tree methods to analyze the influence behaviour of users of recommender systems. Furthermore, [15] proposed a new rating prediction model to estimate the internal factors influence of recommender system users. Indeed, this approach focused on user reliability and popularity in order to alleviate the cold start and data sparsity problems.

From the other hand, some recent research papers have focused on context aware recommender systems. In [16] the researchers propose a leader recognition model based on a social network analysis for identifying opinion leaders in virtual communities of outbound tourism, this is by taking into account the construal influence, content influence, and activity information. [17] proposed to calculate the social and geographical influence of a user on location recommendations by using a kernel density estimation, this is by modelling the geographical influence of locations as individual distributions rather than common distribution for all users. [18] exploited separately the past interactions of users of temporal recommender systems, their social connections, and the similarity between items to measure the influence of users. Then the authors combine the latter features in an interpretable manner. [19] proposed a context-based approach that jointly used current contextual information and social influence to enhance the recommendation of items, based on probabilistic model. The latter approach takes into consideration the context factor such as the time, location of user and weather.

Few studies have focused on using Graph Theory to address the opinion leaders' detection in recommender systems. In this context [20] studied the opinion elicitation and diffusion by proposing an end-to-end Graph-based neural model; while preserving the multi-relations between key opinion leaders and items. [21] presented a weighted graph based approach for leaders extraction using critical vertices whose removal results in the network disconnection.

3 Motivation and Problem Definition

The challenge of opinion leaders' discovering in recommender systems is how to identify the smallest set of users who influence the remaining users by their opinions and ratings. The identified key users may help addressing the cold start item problem by being the first users to rate new items which help predicting the opinions of other users. Additionally, they enable alleviating the cold start user problem by helping new comers to find the appropriate suggestions for them.

Furthermore, most of existing researches on opinion leaders' detection in TRS systems explore single leaders' identification which is not as efficient as parallel detection of the set of opinion leaders as the influence of two users may not be efficient when each of them is detected separately, however if we detect them together they may have a high influence on the TRS network.

Moreover, most proposed approaches for opinion leaders detection focus on choosing the key users in the recommender system without excluding them from the global set of users. Though, the influence of a user is better measured when he is absent which enables evaluating the decrease in network efficiency after his removal.

Graph Theory is a powerful tool for modelling social interactions and detecting key members. Fragmentation and Minimum Vertex Cover have proven to be two efficient graph parameters aiming to extract the most important set of nodes representing the smallest set of users who influence the remaining users' behavior. Indeed, Minimum Vertex Cover (MVC) in a graph is a minimum subset of the vertex set of a given graph G such that every edge in G has at least one endpoint in this set [22] while fragmentation measure intend to select the most efficient set of nodes whose deletion may results in the maximum network division.

To the best of our knowledge, no existing approach has used MVC or Fragmentation to select the key opinion leaders in TRS. Indeed, [21] have used fragmentation to detect key users in non-contextual recommender systems but not in contextual ones, while no existing approach has employed MCV to identify key users neither in non-contextual nor in contextual recommender systems.

A graph parameter is a function $\varphi: G \to R$ where G is the set of finite graphs and R is the set of real numbers. Let φ, ψ be two graph parameters. The parameter φ upper bounds another parameter ψ, if there is some function f such that for every graph g in G it holds that $\psi(G) \leq f(\varphi(G))$ [31].

4 Proposed Methods

4.1 Graph Modeling

In order to identify the set of most important opinion leaders in TRS we model it by a weighted graph representing the social network connecting the users of the system. In this graph, nodes represent users, links represent ties between users and weights of the ties represent the similarity between pairs of users. It should be mentioned that the graph representing the TRS can be complete. The similarity degrees take into account the approximation between the ratings of users on similar points of interests and the users locations as illustrated in the following figure (Fig. 1), where:

- Weights between [0, 0.4] represent bad interactions.
- Weights between [0.5, 0.7] represent average interactions.
- Weights between [0.8, 1] represent good interactions.

In order to calculate the similarity degree between each pair of users, we have used the following similarity measures:

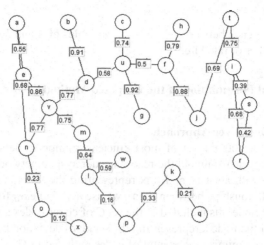

Fig. 1. Graph modeling of the proposed methods

- **Pearson:**
  It is the most used similarity measure in TCRS. It estimates how much the users are correlated by evaluating the deviation of their ratings given on similar items from the average ratings. Pearson similarity coefficient w between user $u1$ and user $u2$ is illustrated in Eq. 1 [23].

$$w(u, v) = \frac{\sum_v (r_{u1,v} - \overline{r_{u1}})(r_{u2,v} - \overline{r_{u2}})}{\sqrt{\sum_v (r_{u1,v} - \overline{r_{u1}})^2} \sqrt{\sum_v (r_{u2,v} - \overline{r_{u2}})^2}} \tag{1}$$

where:
  $r_{u1,v}$ is the rating of user $u1$ on item v
  $v\ r_{u1}$ is the ratings average of user $u1$
  $r_{u2,v}$ is the rating of user $u2$ on item v
  r_{u2} is the ratings average of user $u2$.

- **Kendall:**
  Kendall coefficient is a non-parametric similarity measure that calculates the degree of correlation between pair of users of the recommender system. The closer the similarity value is to 1 the more similar the users are. Kendall coefficient between two users A and B is given Eq. 2 as follows [24]:

$$\text{Kendall}(A, B) = \frac{\text{number of concordant} - \text{number of discordant}}{\frac{1}{2}n(n-1)} \tag{2}$$

where n is the number of objects of concordant and discordant classes.

- **Spearman:**
  Like Kendall coefficient, Spearman is a non-parametric similarity test that calculates how much two users depend to each other [25]. Spearman similarity between two users A and B is given by the following equation (Eq. 3).

$$\text{Spearman}(A, B) = 1 - \frac{6\sum d_i}{n(n^2 - 1)} \tag{3}$$

where,

d_i: is the difference between the ranks of variables of A and B.

n: is the number of rated items.

4.2 Mathematical Formulation of the Proposed Methods

A. Minimum Vertex Cover Approach

Our first approach detects the set of most efficient opinion leaders in TRS based on Minimum Vertex Cover. We model the recommender system network by an undirected graph G = (N, E), with a set of nodes N representing the users and a set of edges E representing the relationships between them, we use a variable α_a for each node i ∈ N. Then, $\alpha_a = 1$ if the node a is included in the set C of cover nodes and 0 otherwise. Let w_a be the weight of the node a representing the standard deviation between the degrees of its ties, we aim to minimize the weight of nodes included in C.

Then, each edges with two vertices (a, b), either a or b or both must be a member of C.

That is $\forall \{a, b\} \in E, \alpha_a + \alpha_b \geq 1$.

The integer program representing the minimum vertex cover (MVC) to select is given by the following formulation (Eq. 4),

$$\min \sum_{a \in N} W_a \alpha_a$$
$$\alpha_a + \alpha_b \geq 1 \forall \{a, b\} \in E \qquad (4)$$
$$\alpha_a \in \{0, 1\} \forall a \in N$$

Also, we aim to maximize the edges weights $W_{a,b}$ of nodes included in the set C of cover nodes to select those nodes that are maximally connected to the remaining nodes as illustrated in formula 5 as follows:

$$\max \sum_{\{a,b\} \in E} W_{a,b} \qquad (5)$$

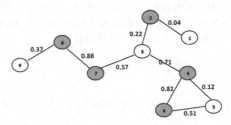

Fig. 2. Opinion leaders detection using minimum cover vertex approach

Figure 2 shows an example of a recommender system graph having 9 nodes and 9 edges, in this example we aim to detect 5 cover nodes. Our Minimum Cover Vertex approach detects the green nodes as most important cover vertices covering more strong ties with higher weights. We have selected node 7 instead of node 3 as it has better weights average with lower standard deviation.

B. Fragmentation Approach

Our second approach uses the cut nodes to detect the set of opinion leaders, its mathematical formulation is given as follows:

Denote an undirected graph $G = (N, E)$ with a set of nodes $N = (i1, i2, ..., i\_n)$ and a set of edges $E = \{ab: a, b \in N\}$. The weights on the edges are given by $W_{a,b} \in [0, 1]$.

We aim to detect a subset of nodes $A \in N$ with the highest weights on the edges, as illustrated in formula 6.

$$arg\ min \sum\nolimits_{i1,i2 \in (N)} Eij(G(N \backslash A)) : |A| \leq S,$$

$$\text{Where:} \begin{cases} E_{i1,i2} = 1, & \text{if } i1 \text{ and } i2 \text{ belongs to the same component of } G(V \backslash A) \\ Otherwise, E_{i1,i2} = 0 \end{cases} \quad (6)$$

The objective is to identify a subset of nodes $A \subset N$ where $|A|$ is smaller or equal to a value S, and whose removal generates the minimum set of edges Eij included in the subgraph $G(N \backslash A)$.

The subset of nodes A must also have the maximum weights average $W_{a,b}$ as depicted in Eq. 7:

$$\max \sum\nolimits_{\{a,b\} \in N} W_{a,b} \quad (7)$$

In addition, the standard deviation of the edges weights $W_{a,b}$ of cut nodes is greater than or equal to a threshold value β.

It should be mentioned that the weights on the edges have been normalized using *sigmoid function* that varies between [0, 1]. The latter equation is given in Eq. 8.

$$y = \frac{1}{1 + \varepsilon^{-x}} \quad (8)$$

where x is the original value and y is the normalized value.

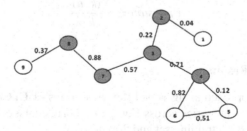

Fig. 3. Opinion leaders detection using fragmentation approach

Figure 3 depicts an example of the recommender system graph shown in Fig. 2, in this graph we can detect 5 cut nodes (green nodes) as their deletion results in fragmentation of the graph into different connected component. Our objective is two select the most powerful set including three nodes, thus, our Fragmentation approach identifies node 3, node 7 and node 4 as a set of most important cut nodes.

Indeed, deletion of node 3 generates three fragments $\{1, 2\}$, $\{4, 5, 6\}$ as well as $\{7, 8, 9\}$, deletion of node 7 gives two fragments $\{8, 9\}$ and $\{1, 2, 3, 4, 5, 6\}$, and deletion of node 4 creates two fragments $\{5, 6\}$ and $\{1, 2, 3, 7, 8, 9\}$. It should be noted that we have taken the cut nodes having the significant similarity scores and whose deletion results in important separated components.

5   Results and Discussion

5.1   Evaluation Measures

In order to evaluate the performance of our Vertex Cover and Fragmentation approaches and validate their efficiency, we have used Mean Absolute Error and Precision evaluation metrics.

A. Mean Absolute Error (MAE)
This measure evaluates the average of the absolute difference between the ratings of users and their computed predictions. MAE is a negative oriented coefficient, thus, lower values of error are better. Equation 9 gives MAE measurement [26]

$$MAE = \frac{\sum_{\{i,j\}} |p_{i,j} - r_{i:j}|}{n} \tag{9}$$

where $r$max and $r$min are the upper and lower bounds of the ratings.

B. Precision
It is a prediction measurement that estimates the fraction of relevant items recommendations for user u among the k generated recommendations. It is given as follows (Eq. 10), where $N(k, u)$ is the set of relevant items in the test set which is included in the top-k of the ranking list [27]

$$Precision = \frac{N(k, u)}{k} \tag{10}$$

5.2   Experimental Results

We have tested our approach using Travel Review Ratings - UCI Google review ratings on attractions with 24 categories across Europe [28]. This dataset includes 5456 users and was divided into 80% training set and 20% test set.

A. MAE Results
The results of MAE values when deleting opinion leaders is shown in Figs. 4, 5, 6 and 7 bellow.

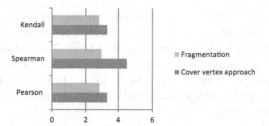

Fig. 4. TOP-10 Opinion leaders using Fragmentation and Minimum Cover Vertex approaches

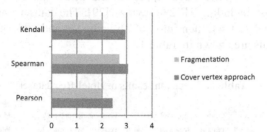

Fig. 5. TOP-20 Opinion leaders using Fragmentation and Minimum Cover Vertex approaches

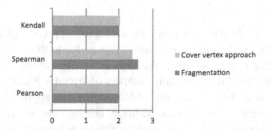

Fig. 6. TOP-30 Opinion leaders using Fragmentation and Minimum Cover Vertex approaches

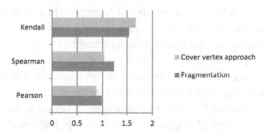

Fig. 7. TOP-40 Opinion leaders using Fragmentation and Minimum Cover Vertex approaches

From Figs. 4 and Fig. 5 it can be clearly seen that Fragmentation approach performs much better than Minimum Cover Vertex approach when the number of opinion leaders is lower (TOP 10 and TOP 20 opinion leaders). This is due to the effect of connections in the recommender system graph; since cut nodes connects the recommender communities and their absence creates separated components; where cover nodes links the graph ties without linking a considerable number of connected components. However, in TOP 30 and TOP 40 opinion leaders results (Fig. 6 and Fig. 7) we observe that Minimum Cover Vertex approach outperforms Fragmentation approach, this is because the nodes representing the opinion leaders cover more sets of ties. In addition, Pearson coefficient has given better results than Spearman and Kendall this is because it evaluates the deviation of all given ratings.

B. Precision Results

We have tested both Fragmentation Approach and Minimum Cover vertex approach, on Hotel-Rec dataset including 717280 reviews [29]. The dataset was divided into 85% training set and 15% test set, then into 80% training set and 20% test set respectively. The precision results are shown in Table 1.

Table 1. Precision results of Hotel-Rec dataset

Test corpus	Centrality approach [30]			Connectors approach [21]			Fragmentation approach			Minimum cover vertex approach		
	Pearson	Spearman	Kendall	Pearson	Spearman	Kendall	Pearson	Spearman	Kendall	Pearson	Spearman	Kendall
15%	72,73%	45,45%	57,14%	82,02%	55,87%	83,47%	**85,72%**	57,14%	68,89%	**84,66%**	51,37%	66,42%
20%	88,24%	82,35%	76,92%	84,44%	63,05%	79,38%	90,48%	66,67%	**97,17%**	**95,65%**	59,97%	92,14%

From Table 1 we observe that Our Fragmentation approach gives better precision results than centrality and connectors results. This is due to the cut points' efficiency as they link the connected communities of the TRS, while Centrality technique detects those nodes who are connected to maximum numbers of friend without taking into account the fragmentation of users communities. Also, connectors approach does not take into account the distribution of weights. In addition, our fragmentation approach identifies simultaneously sets with the smallest number of nodes with the most efficient similarity scores.

Furthermore, Minimum Cover vertex approach outperforms centrality method because it cover more important ties in the TRS graph while centrality focuses on the number of incoming and outgoing ties without taking into account the covered paths. Also, Pearson and Kendal are the best similarity measures as they are appropriate to the different ratings scales of users of Hotel-Rec dataset.

Table 2 illustrates a comparison between our approach and existing approaches for opinion leaders identification in recommender systems in terms of features.

Based on the structure of the recommender system network, we can see in Table 2, that our approach uses the graph parameters to efficiently and simultaneously detect the set of most important key opinion leaders. The other approaches have used basic graph concepts such as centrality and density without integrating graph parameters.

Furthermore, our approach has represented the recommender system by weighted graphs to measure the proximity between users' context information and their opinions. Indeed, weighted graphs allow providing more useful information about relationships and similarities between users. In addition, [16] and [19] approaches have focused on directed graphs that interpret relationships in both directions. The performance of our

Table 2. Comparison between our approach and existing approaches in terms of features

Approach	Graph parameters	Simultaneous detection of opinion leaders	Weighted graph	Direct graph	Rating	Context information
[16]	−	−	+	+	−	−
[19]	−	−	+	+	+	+
[20]	−	+	+	−	+	−
[21]	+	+	+	−	+	−
Our approach	+	+	+	−	+	+

approach was also guaranteed by the use of both ratings of users and their context aware information to generate the most appropriate suggestions of items for travellers.

6 Conclusion

In this paper, we have presented two approaches for detecting opinion leaders in travel recommender systems. Our methods were based on minimum cover vertex and fragmentation using cut nodes. We also considered the similarity score between users using their ratings as well as their contextual information. Experimental results have shown that cut nodes approach was more efficient than cover vertices approach due to their ability of connecting the recommender system communities. Indeed, Fragmentation approach achieved 97,17% of precision.

As future works, we will use other graph parameters for detecting opinion leaders in recommender system graphs. Another interesting trend is the use of deep learning algorithms in order to enhance the results with tests on larger datasets. Our approach can also be enriched by integrating bidirectional users' interactions into the prediction process.

References

1. Liu, Q., Ma, H., Chen, E., Xiong, H.: A survey of context-aware mobile recommendations. Int. J. Inf. Technol. Decis. Mak. **12**(1), 139–172 (2013)
2. Árnason, J.I., Jepsen, J., Koudal, A., Schmidt, M.R., Serafin, S.: Volvo intelligent news: a context aware multi modal proactive recommender system for in-vehicle use. Pervasive Mob. Comput. J. **14**, 95–111 (2014)
3. Hana, J., Schmidtke, H.R., Xie, X., Woo, W.: Adaptive content recommendation for mobile users: ordering recommendations using a hierarchical context model with granularity. Pervasive Mob. Comput. J. **13**, 85–98 (2014)
4. Patra, B.K., Launonen, R., Ollikainen, V., Nandi, S.: A new similarity measure using Bhattacharyya coefficient for collaborative filtering in sparse data. Knowl. Based Syst. **82**, 163–177 (2015)

5. Bobadilla, J., Ortega, F., Hernando, A., Gutiérrez, A.: Recommender systems survey. Knowl.-Based Syst. **46**, 109–132 (2013)
6. Badaro, G., Hajj, H., El-Hajj, W., Nachman, L.: A hybrid approach with collaborative filtering for recommender systems. In: 9th International Wireless Communications and Mobile Computing Conference (IWCMC), Sardinia, pp. 349–354 (2013)
7. Braunhofer, M.: Hybrid solution of the cold-start problem in context-aware recommender systems. In: Dimitrova, V., Kuflik, T., Chin, D., Ricci, F., Dolog, P., Houben, G.-J. (eds.) UMAP 2014. LNCS, vol. 8538, pp. 484–489. Springer, Cham (2014). https://doi.org/10.1007/978-3-319-08786-3_44
8. Sheridan, P., Onsjö, M., Becerra, C., Jimenez, S., Dueñas, G.: An ontology-based recommender system with an application to the Star Trek television franchise. Future Internet **11**(9), 1–23 (2019)
9. Nguyen, Q., Huynh, L.N.T., Le, T.P., Chung, T.: Ontology-based recommender system for sport events. In: Lee, S., Ismail, R., Choo, H. (eds.) IMCOM 2019. AISC, vol. 935, pp. 870–885. Springer, Cham (2019). https://doi.org/10.1007/978-3-030-19063-7_69
10. Das, J., Mukherjee, P., Majumder, S., Gupta, P.: Clustering-based recommender system using principles of voting theory. In: International Conference on Contemporary Computing and Informatics (IC3I), pp. 230–235 (2014)
11. Shulman, E., Wolf, L.: Meta decision trees for explainable recommendation systems. In: Machine Learning (2020)
12. Shriver, D., Elbaum, S., Dwyer, M.B., Rosenblum, D.S.: Evaluating recommender system stability with influence-guided fuzzing. In: The Thirty-Third AAAI Conference on Artificial Intelligence (AAAI 2019), pp. 4934–4942 (2019)
13. Eskandanian, F., Sonboli, N., Mobasher, B.: Power of the few: analyzing the impact of influential users in collaborative recommender systems. In: Social and Information Networks. ACM Publisher (2019)
14. Morid, M.A., Shajari, M., Golpayegani, A.H.: Who are the most influential users in a recommender system? In: The 13th International Conference on Electronic Commerce, pp. 1–5 (2011)
15. Shi, W., Wang, L., Qin, J.: Extracting user influence from ratings and trust for rating prediction in recommendations. Scientific Reports IF4.379 (2020)
16. Yang, J., Zhang, Y., Liu, L.: Identifying opinion leaders in virtual travel community based on social network analysis. In: Nah, F.-H., Siau, K. (eds.) HCII 2019. LNCS, vol. 11589, pp. 276–294. Springer, Cham (2019). https://doi.org/10.1007/978-3-030-22338-0_23
17. Zhang, J., Chow, C.Y.: GSLR: personalized geo-social location recommendation - a kernel density estimation approach. In: Proceedings of the 21st ACM SIGSPATIAL International Conference on Advances (2013)
18. Narang, K., Song, Y., Schwing, A., Sundaram, H.: FuseRec: fusing user and item homophily modeling with temporal recommender systems. Data Min. Knowl. Disc. **35**(3), 837–862 (2021). https://doi.org/10.1007/s10618-021-00738-8
19. Bambia, M.: Jointly integrating current context and social influence for improving recommendation. Ph.D. thesis, University of Paul Sabatier, Toulouse III (2017)
20. Wang, J., Ding, K., Zhu, Z., Zhang, Y., Caverlee, J.: Key opinion leaders in recommendation systems: opinion elicitation and diffusion. In: The 13th International Conference on Web Search and Data Mining, WSDM 2020, Texas (2020)
21. Chekkai, N., et al.: CSCF: clustering based-approach for social collaborative filtering. In: 2017 First International Conference on Embedded & Distributed Systems (EDiS), Oran, pp. 1–6 (2017)
22. Gu, J., Guo, P.: PEAVC: an improved minimum vertex cover solver for massive sparse graphs. Eng. Appl. Artif. Intell. **104** (2021)

23. Ekstrand, M.: Similarity Functions for User-User Collaborative Filtering (2013). https://gro uplens.org/blog/%20similarity-functions-for-useruser-collaborative-filtering/

24. Arsan, T., Koksal, E., Bozkus, Z.: Comparison of collaborative filtering algorithms with various similarity measures for movie recommendation. Int. J. Comput. Sci. Eng. Appl. (IJCSEA) **6**(3), 1–20 (2016)

25. Yu, Y., Shanfeng, Z., Xinmeng, C.: Collaborative filtering algorithms based on Kendall correlation in recommender systems. Wuhan Univ. J. Nat. Sci. **11**(5), 1086–1090 (2006)

26. Sarwar, B., Karypis, G., Konstan, J., Riedl, J.: Item-based collaborative filtering recommendation algorithms. In: Proceedings of the 10th International Conference on World Wide Web, WWW 2001, pp. 285–295 (2001)

27. Yang, X., Guo, Y., Liu, Y., Steck, H.: A survey of collaborative filtering based social. Comput. Commun. J. **41**, 1–10 (2012)

28. Travel Review Ratings. https://www.kaggle.com/ishbhms/travel-review-ratings. Accessed 04 Apr 2022

29. Hotel-Rec Dataset 8. https://www.kaggle.com/hariwu1995/hotelrec-dataset-8. Accessed 04 Apr 2022

30. Jain, L., Katarya, R., Sachdeva, S.: Role of opinion leader for the diffusion of products using epidemic model in online social network. In: The 2019 Twelfth International Conference on Contemporary Computing (IC3), pp. 1–6. IEEE (2019)

31. Sorge, M., et al.: The graph parameter hierarchy. https://citeseerx.ist.psu.edu/viewdoc/sum mary?doi=10.1.1.412.4918. Accessed 14 Aug 2022

DETECT Workshop: Modeling, Verification and Testing of Dependable Critical Systems

Revisiting Ontology Evolution Patterns
A Formal xDSL Approach

Akram Idani[1]([✉]), Rim Djedidi[2], and German Vega[1]

[1] Univ. Grenoble Alpes, Grenoble INP, CNRS, LIG, 38000 Grenoble, France
{Akram.Idani,German.Vega}@univ-grenoble-alpes.fr
[2] LIMICS - INSERM Laboratory, Sorbonne Paris Nord University, Bobigny, France
rim.jedidi@univ-paris13.fr

Abstract. Mixing formal methods (such as B) and ontology description languages is a promising approach to develop reliable, safe and secure systems. In fact, both techniques have demonstrated their strengths to handle ambiguity, inconsistency and incompleteness of requirements. However, domain knowledge evolves continuously throughout the application life-cycle and hence different change requirements must be addressed. Several problems have to be managed in the ontology evolution process particularly change impact analysis and resolution. Among existing works, pattern-driven techniques have been proposed to provide guidance during the ontology evolution so that it remains consistent. The underlying process is often informal and mostly methodological; and consequently automation efforts are still required. In this paper, we propose a lightweight pattern-driven approach built on an executable formal definition of the Ontology Web Language (OWL), that we call xOWL. In order to ensure the execution, the debugging and the correctness of existing patterns, we instrumented xOWL using the B method and the Meeduse language workbench.

Keywords: Ontologies · OWL · B Method · MDD · DSLs

1 Introduction

Knowledge engineering may be highly critical when it is intended to make important decisions such as provide a medical advice [15] or deal with a security issue [9]. For this reason several research works [2,3,8,16] propose to embed a formal method (*e.g.* Event-B, Maude) within well-established paradigms for domain knowledge representation. Indeed, formal methods are nowadays the most rigorous way to produce correct artifacts thanks to the availability of automated reasoning tools such as provers and model-checkers. By mixing knowledge description languages and formal approaches, one can benefit from their cross contributions to ensure the precision as well as the readability of the requirements, which is useful for the verification and validation (V&V) activities. In this work we are interested by two major kinds of approaches applied to the Ontology Web Language OWL[1]: (1) pattern-driven approaches [5–7,18] whose

---

[1] http://www.w3.org/TR/owl2-overview/.

© The Author(s), under exclusive license to Springer Nature Switzerland AG 2022
P. Fournier-Viger et al. (Eds.): MEDI 2022, CCIS 1751, pp. 165–178, 2022.
https://doi.org/10.1007/978-3-031-23119-3_12

objective is to provide some guidance during the ontology evolution so that it remains consistent all along its life-cycle; and (2) translational approaches [3,8] in which the ontology is translated into a mathematically defined language that is assisted by theorem provers and/or constraint solvers.

Every approach has its strengths and weakness. In the former, ontology evolution follows a well-established pattern-oriented process dealing with three major concerns: change, inconsistency and resolution. However, the underlying process remains somehow informal and mostly methodological. In order to make these approaches more effective, automation efforts are required not only to provide a tool support, but also to deal with the correct application of the provided patterns. In the latter, the translation from a given ontology to a formal language applies a model-to-model transformation that is expected to extract an equivalent formal model. The advantage is that obviously the target language allows one to remedy the lack of verification tools. However, these approaches deal only with the correctness issues and do not address the ontology evolution problem. Furthermore, the model-to-model transformation they provide is often hard-coded and difficult to extend or adapt for a better coverage of ontologies.

To circumvent these shortcomings, we propose a lightweight development approach built on a formal model-driven executable domain-specific language (xDSL), that we call in the remainder xOWL. The aim of an xDSL is not only to represent the structural features of a system, but also its behaviour. Considering that OWL is a DSL (dedicated to ontologies), our proposal is to rethink the evolution patterns by means of execution semantics that apply the expected changes to a given ontology. In order to ensure automation as well as correctness, we instrumented xOWL in Meeduse [12], the only existing language workbench today that favours both formal reasoning – via theorem proving – and the execution of the DSL – via animation and model-checking. Meeduse applies B method [1] to define the semantics of a DSL and embeds ProB [13], a powerful animator, model-checker and constraint-solver of the B method.

This paper is structured as follows: Sect. 2 discusses the spectrum of our work and presents the contributions regarding the state of the art. Section 3 provides an overall view about our approach. In Sect. 4 we show how the semantic embedding of B and OWL is done and illustrate it on the SubClassOf pattern. Finally, Sect. 5 draws the conclusion and perspectives of this work.

2   Pattern-Driven Change Management

Our work focuses on issues related to pattern-driven ontology change management and especially on consistency maintenance [6]. These patterns have been proposed as a solution looking for invariances that repeatedly appear when evolving ontologies. They are approved by the ontology community and shared in the ontology design patterns (ODP[2]) portal.

They help determining the inconsistencies that could be potentially caused by a type of change and the alternatives that may resolve a kind of inconsistency.

---

[2] http://ontologydesignpatterns.org/.

We address two consistency levels: structural level [11] (conformance with the ontology specification language) and logical level [10] (assertions do not present contradictions). A logical inconsistency is more difficult to detect because, on the one hand it requires the usage of a mathematical language assisted by reasoners, and on the other hand, it can be localized at various abstraction levels (terminology and concrete entities of a domain). The approach of this paper addresses both abstraction levels, however, for the sake of readability we will focus on the terminology level.

2.1 Example

In order to illustrate our approach step-by-step, we apply it to the SubClassOf pattern defined in [6], and we reuse the underlying OWL axioms:

$$Animal \sqsubseteq Fauna\_Flora, \; Plant \sqsubseteq Fauna\_Flora,$$

$$Carnivorous\_Plant \sqsubseteq Plant, \; Plant \sqsubseteq \neg Animal$$

A change defining Carnivorous_Plant class as a sub-class of class Animal ($Carnivorous\_Plant \sqsubseteq Animal$) causes a disjointness inconsistency. In fact, a carnivorous plant is supposed to be a plant and also an animal. However, the concepts are disjoint. To solve this inconsistency, the SubClassOf pattern suggests two alternatives (cf. Fig. 1): hybrid sub-class (Fig. 1a) and hybrid super-class (Fig. 1b).

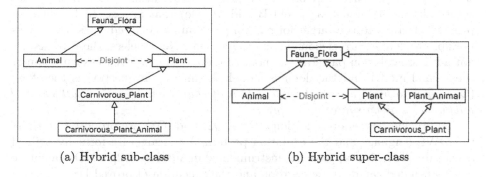

(a) Hybrid sub-class   (b) Hybrid super-class

Fig. 1. Two alternatives to solve a sub-class inconsistency.

The hybrid class Plant_Animal is based on the definition of the two disjoint classes involved in the inconsistency: Animal and Plant. Both alternatives solve the disjointness inconsistency and preserve existing knowledge. As a logical inconsistency resolution, the two alternatives do not take into account domain specific knowledge. The ontology engineer and/or domain expert can then choose the most suitable alternative according to the domain semantics and the ontology usage context. This example shows that an evolution pattern deals with three concerns: (*i*) changes to be applied, (*ii*) logical properties to be preserved and (*iii*) resolution alternatives to keep the ontology consistent.

2.2   Discussion and Related Works

There exist several works that addressed the ontology evolution problem. In [7] a comparative review based on general and specific characteristics and supported features (such as consistency maintenance) for ontology evolution is provided. A more recent survey is proposed in [18]; the authors adopt a process-centric view of ontology evolution that includes change validation step. Based on these surveys, a common question appears: (Q1) *"how to preserve the consistency of the ontology during its evolution?"*.

Few approaches provide established change management operations (patterns) that can be applied during the evolution process. We refer to the work of Djedidi and Aufaure [6] on which we build our proposal, and to that of Castano et al., [5] where the authors propose four evolution patterns covering population (new instances) and enrichment (new concepts, relations, properties). The latter work takes into account structural and logical consistency validation. But, it is limited to inconsistency detection. The added-value of [6] is that it provides various alternative solutions in case of inconsistency. However, a major question remains open: (Q2) *"how to automate a pattern-driven change management?"*.

Regarding the verification and validation (V&V) activities, often reasoners (such as HermiT, Fact++, Pellet) are used to pinpoint inconsistencies. The drawback of this approach is that the inconsistency is detected after it happens. In practice, a more pragmatic approach would be the application of a change without introducing any flaw − since there exist some alternative solutions. Approaches such as [2,3,8,16] propose the translation of the ontology into a formal language (such as Event-B and Maude) that is assisted by theorem proving. Thanks to the target formal language and associated tools it becomes possible to define correct domain specific actions. Nonetheless, these works did not address evolution patterns and none of them offered a way to execute jointly the formal model and the domain model. A third open question is therefore: (Q3) *"how to build a correct specification of change management patterns and apply it to a given source ontology?"*.

To provide answers to questions (Q1), (Q2) and (Q3), we propose to rethink the OWL language and the evolution patterns by means of a formally defined executable DSL. Our approach is instrumented in Meeduse [12], a MDE language workbench dedicated to the creation and the execution of proved DSLs.

3   The xOWL Approach

In the literature, one can find two kinds of formal modelling to which an ontology can be mapped [2]: shallow modelling and deep modelling. In a shallow modelling, the ontology is directly translated into a formal model without keeping trace of its semantics. This approach is domain-centric and does not favour the ontology evolution since changes may impact the data structures that are initially generated. In a deep modelling, the ontology and its underlying semantic domain are mapped together to the target formal model. The resulting data structures define the various concerns of the ontology description language, and hence the ontology evolution do not have any impact on these data structures. In this approach, a given ontology is seen as a data-set that valuates the data structures issued from the ontology language. This paper applies a deep modeling approach built on the formal B method, providing a mathematical specification of the evolution patterns and a formal reasoning about their correctness. Our patterns are managed by Meeduse [12], which offers several useful features such as: verification, debugging and execution.

3.1 Architecture

The overall architecture of the xOWL approach is presented in Fig. 2. The semantic layer builds on three artifacts: OWL grammar, formal B model and evolution patterns.

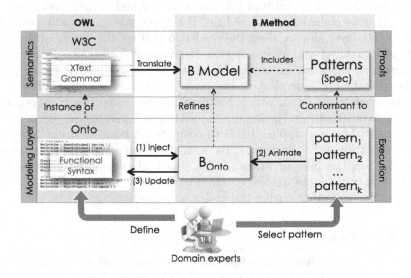

Fig. 2. Semantic embedding of B and OWL.

1. The OWL grammar: defined using Xtext [4], a well-known DSL builder. The tool allows one to define textual languages using LL(*) grammars and generate a full infrastructure, including an ANTLR parser API with a type-checker and auto-completion facilities. We have encoded the OWL2 functional syntax[3] defined by the W3C consortium.
2. A formal B model: automatically extracted by Meeduse from the Xtext grammar. In the remainder, we call this model functional because it refers to the functional syntax of OWL. It is about 1632 lines and features all B data structures that are related to the grammar rules. It also provides several B operations to ensure basic actions such as getters, setters, constructors and destructors. The advantage of this formal specification is that it is proved

---

[3] https://www.w3.org/TR/owl2-syntax/.

correct: 721 proof obligations were generated by the AtelierB prover, among them 685 were proved automatically and 36 interactively. These proofs attest that the provided basic operations preserve the structural consistency of the ontology and do not produce syntactic errors if applied.

3. The evolution patterns: defined with a catalog of high-level B specifications in which the logical properties correspond to B invariants and the desired changes to B operations. These specifications apply the functional B model in order to ensure the ontology evolution. Since the structural properties of the OWL language are preserved by the functional model, a pattern specification includes only the logical properties that must be respected during the evolution. The proof of correctness means that the pattern preserves also the logical properties that are involved in the pattern definition.

Regarding the modeling layer, it follows three steps (inject, animate, update) and benefits from the ProB tool [13], an animator and model-checker of the B method that is embedded in Meeduse. For space reason we will not discuss in detail this layer and we refer the reader to [12] for the execution feature of Meeduse. Roughly speaking, starting from a given ontology (called Onto in Fig. 2) conforming to the Xtext grammar, Meeduse creates a valued B model (B_{Onto}) that is semantically equivalent to the input ontology. B_{Onto} is a refinement of the functional B model in which Meeduse injects valuations to populate the various B data structures. Having this valued B model and the catalog of patterns written in B, Meeduse applies ProB to compute all possible instances of these patterns ($pattern_i$, ..., $pattern_k$). By animating interactively these instances, ProB modifies the internal state of B_{Onto} and Meeduse translates back this modification to the input ontology so that both models (Onto and B_{Onto}) remain equivalent to each other during the evolution process.

3.2   Illustration

Figure 3 is a screen-shot of Meeduse while running xOWL with a formal specification of the SubClassOf pattern with logical inconsistency. The top side of the figure represents our simple ontology example presented in both textual and graphical syntax. The textual editor (1) is generated by Xtext from the OWL grammar and the graphical editor (2) is designed via Sirius [17]. Note that the synchronisation between the tools is transparent [14] and does not require any additional implementation effort.

The other views of Fig. 3 are provided by Meeduse to ensure execution, verification and debugging: the execution history view (3) allows an omniscient debugging, which is useful to go backward in time and manage the execution traces; the execution view (4) provides the pattern instances by means of valued B operations that can be applied to the current model; the command-line console (5) is for interactive debugging; and finally, the state view (6) computes the various valuations of the B data structures and informs the user whether the invariant properties are preserved or not.

Figure 3 presents a straightforward application of the SubClassOf pattern with inconsistency. The underlying specification builds on three B operations:

selectClasses, createSubClassOf and addSubClassOfAxiom. The first operation allows one to select the OWL classes between which a sub-class relation will
be created. In the current model the tool computes four possible applications of
this operation (cf. Execution view(4)) and states that all of them are not_safe.
The other operations (*i.e.* createSubClassOf and addSubClassOfAxiom) apply
the required modifications to the model: creation of a new sub-class axiom,
and insertion of the axiom within the ontology. The execution of this pattern to
classes Carnivorous_Plant and Animal produces the ontology of Fig. 4 in which
the resulting inconsistency is highlighted via an annotation.

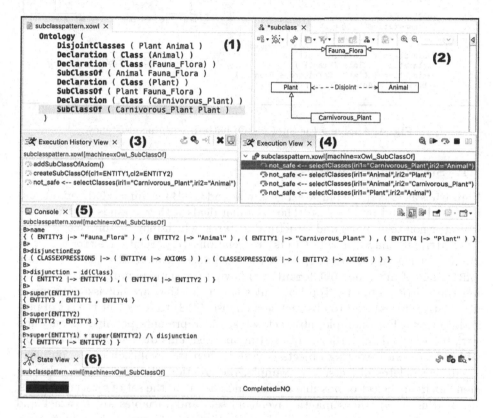

Fig. 3. xOWL instrumented in Meeduse.

4   Semantic Integration of OWL and B

In the previous section we presented and illustrated an overview of the xOWL
approach. The originality of xOWL is that it combines, in one unifying framework, two interesting paradigms: formal methods and executable DSLs. Furthermore, xOWL is a lightweight approach because, in addition to the Xtext grammar
(that has been already done) and possibly the graphical syntax (which may be
more convenient than the textual syntax), the development effort is limited to

the specification of the evolution patterns and their verification and validation. Indeed, all the other artifacts of Fig. 2 are managed by the Meeduse language workbench. In this section we show how the semantic integration of OWL and B is done, and discuss a formal definition of the SubClassOf pattern.

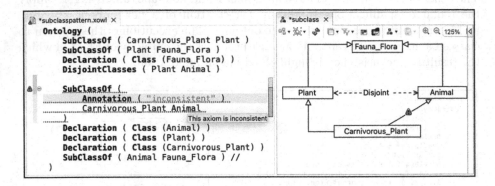

Fig. 4. Applying the SubClassOf pattern with inconsistency.

4.1 Structural Properties

Figure 5 is an excerpt of our Xtext grammar and the B machine (xOWLSemantics) that is generated by Meeduse. The grammar deals with two kinds of structures: rules (or non-terminals) and structural features, which may be simple attributes (*e.g.* attribute **name** of rule **Class**) or relations (*e.g.* relation **axioms** of rule **xOWL** referring to rule **Axiom**). All these structures lead to B variables (clause **VARIABLES** of machine **xOWLSemantics**) having the same naming and representing the existing objects. Top level rules (*i.e.* rules that are not issued from an alternatives rules) lead to abstract sets (clause **SETS**) to represent the set of possible objects. For example, abstract set **AXIOM** represents possible axioms and variable **Axiom** defines the set of existing axioms. This distinction between possible objects and existing objects is useful when the evolution of the ontology creates or deletes objects. For example, the creation of a new axiom takes an element from the set of possible axioms and adds it to the set of existing axioms.

At this stage (the semantic layer), all sets and variables are abstract and not related to valuations because the objective is to focus on the specification principles, not yet on the execution. In addition to these B data structures, Meeduse generates the structural invariants as shown in Fig. 6.

In front of the invariants we refer to the lines of the grammar from which they are produced. The extraction of these invariants follows four major principles:

- Top level rules: lead to set inclusion between possible and existing objects (*e.g. Axiom $\subseteq$ AXIOM*).
- Alternatives rules: also lead to set inclusion between the variables that represent the rule and its possible alternatives. For example, rule **Axiom** (**line 7**) is a choice between a declaration or a class axiom. Hence, the resulting invariant is: *Declaration $\subseteq$ Axiom $\wedge$ ClassAxiom $\subseteq$ Axiom.*

```
 1 grammar fr.lig.vasco.XOwl with org.eclipse.xtext.cc
 2⊖xOwl:{xOwl}
 3     'Ontology' '('
 4         (axioms+=Axiom)*
 5     ')';
 6
 7 Axiom: ( Declaration | ClassAxiom );
 8
 9 ClassAxiom: ( SubClassOf | DisjointClasses ) ;
10
11⊖DisjointClasses:
12     'DisjointClasses' '('
13         (axiomAnnotations+=Annotation)*
14         (disjointClassExp+=ClassExpression)+
15     ')' ;
16
17⊖SubClassOf:
18     'SubClassOf' '('
19         (axiomAnnotations+=Annotation)*
20         subClassExp=ClassExpression
21         superClassExp=ClassExpression
22     ')' ;
23
24⊖Annotation:
25     'Annotation' '(' annotationValue=STRING ')';
26
27 Declaration: 'Declaration' '(' entity=Entity ')' ;
28
29 Entity: 'Class' '(' Class ')' ;
30
31 ClassExpression: class=[Class] ;
32
33 Class: name=ID ;
```

MACHINE
  xOWLSemantics
SETS
  XOWL ; *AXIOM* ; *ENTITY* ;
  ANNOTATION ; *CLASSEXPRESSION*

VARIABLES

/* grammar non-terminals */	/* grammar attributes */
XOwl,	*axioms,*
Axiom,	*axiomAnnotations,*
Annotation,	*class,*
ClassExpression,	*disjointClassExp,*
Entity,	*subClassExp,*
ClassAxiom,	*superClassExp,*
Class,	*entity,*
Declaration,	*annotationValue,*
DisjointClasses,	*Entity_name*
SubClassOf,	

Fig. 5. Excerpt of our xOWL grammar and its B data structures.

– Rule attributes and relations: lead to specializations of functional relations
  depending on their arities (single/multi valued) and character (mandatory
  or optional). For example, attribute **name** of rule **Class** is mandatory and

INVARIANT

/* Top level rules */

$XOwl \subseteq XOWL \wedge Axiom \subseteq AXIOM \wedge Entity \subseteq ENTITY \wedge$

$Annotation \subseteq ANNOTATION \wedge$

$ClassExpression \subseteq CLASSEXPRESSION \wedge$

/* Alternatives rules */

$Declaration \subseteq Axiom \wedge ClassAxiom \subseteq Axiom$	line 7
$\wedge \, SubClassOf \subseteq ClassAxiom \wedge DisjointClasses \subseteq ClassAxiom$	line 9
$\wedge \, Class \subseteq Entity$	line 29

/* Lists */

$\wedge \, axioms \in Axiom \twoheadrightarrow XOwl$	line 4
$\wedge \, axiomAnnotations \in Annotation \twoheadrightarrow ClassAxiom$	line 13
$\wedge \, disjointClassExp \in ClassExpression \twoheadrightarrow DisjointClasses$	line 14

/* Single values */

$\wedge \, class \in ClassExpression \twoheadrightarrow Class$	line 31
$\wedge \, subClassExp \in SubClassOf \rightarrowtail ClassExpression$	line 20
$\wedge \, superClassExp \in SubClassOf \rightarrowtail ClassExpression$	line 21
$\wedge \, entity \in Declaration \rightarrowtail\!\!\!\rightarrow Entity$	line 27
$\wedge \, annotationValue \in Annotation \rightarrow STRING$	line 25
$\wedge \, Entity\_name \in Class \rightarrow STRING$	line 33

/* Language restrictions */

$\wedge \, \forall \, cc \cdot (cc \in \mathbf{ran}(disjointClassExp) \Rightarrow \mathbf{card}(disjointClassExp^{-1}[\{cc\}]) \geq 2)$

$\wedge \, Declaration \cap ClassAxiom = \emptyset \wedge Declaration \cup ClassAxiom = Axiom$

$\wedge \, SubClassOf \cap DisjointClasses = \emptyset$

$\wedge \, SubClassOf \cup DisjointClasses = ClassAxiom$

Fig. 6. Structural B invariants of xOWL.

single-valued; it is translated into a total function ($\rightarrow$) from variable `Class` to type `STRING`. Relation `axioms` means that an ontology may have zero or many axioms and an axiom belongs to at the most one ontology; it is then translated into a partial function ($\twoheadrightarrow$) from variable `Axiom` to variable `XOwl`.

 - Language restrictions: refer to additional invariants that restrict the usage of the language. For example, the first invariant in section /* **Language restrictions** */ of Fig. 6 means that a disjoint class axiom links two class expressions at least. The grammar (`Line 14.`) applies a one-to-many relation, but the usage of the OWL language requires a two-to-many relation. The other invariants are issued from alternatives rules. For example, $Declaration \cap ClassAxiom = \emptyset \wedge Declaration \cup ClassAxiom = Axiom$ means that an axiom is either a declaration or a class axiom.

The operational part of machine `xOWLSemantics` provides 122 B operations, which allows the developer to apply modifications to a given model without violating the above structural properties of the OWL language. In our work, these operations are applied in the formal specification of the evolution patterns.

4.2   Logical Properties

As the case study of this paper is that of the SubClassOf pattern, we mainly focus on the properties that this pattern must preserve, in addition to the structural ones. The development of other patterns follows the same process: (1) enrich our catalog of logical properties using B definitions and invariants, (2) specify the pattern, and (3) prove its correctness by theorem proving or model-checking. Figure 7 is the header part of machine `SubClassOfPattern`: it includes machine `xOWLSemantics` in order to reuse its data structures and apply its operations. The figure shows the B definitions and invariants of the pattern's logical properties. These definitions compute inheritance cycles and inconsistent classes, and the invariant means that both sets must be empty.

MACHINE *SubClassOfPattern*
INCLUDES *xOWLSemantics*
DEFINITIONS
  $inheritance == (subClassExp ; class)^{-1} ; (superClassExp ; class)$;
  $disjunctionExp == class \otimes disjointClassExp$;
  $disjunction == (\mathbf{ran}(disjunctionExp) ; \mathbf{ran}(disjunctionExp)^{-1}) - \mathbf{id}(Class)$;
  $super(cls) == \mathbf{closure}(inheritance)[cls]$;
  $cycles == \mathbf{closure1}(inheritance) \cap \mathbf{id}(Class)$;
  $inconsistency == \{ cl \mid cl \in Class \wedge disjunction[super(\{cl\})] \cap super(\{cl\}) \neq \emptyset \}$
INVARIANT
  $cycles = \emptyset \wedge inconsistency = \emptyset$

Fig. 7. B definitions and invariants.

In order to illustrate this specification we execute it on our simple example and we use the interactive debugger of Meeduse (cf. Fig. 8). We limit our illustration here to definition *inheritance*. Valuations in capital letters are object identifiers that are extracted by Meeduse from the model. Every entry (starting with B>) is a B expression that is evaluated by the debugger. The illustration shows the evaluation results for every sub-expression of the definition. The idea is to compute, based on the xOWL grammar and its functional B specification, all class couples that are related with an inheritance link.

Definition *cycles* applies a non-reflexive transitive closure over set *inheritance* and computes the intersection with the identity of set *Class*. This intersection is not empty if a cycle exists. Definition *Super(cls)*, when applied to a set of classes (*i.e. cls*), computes all the super classes of this set. It is a reflexive transitive closure, which means that set *cls* is included in the result. Having this, definition *inconsistency* computes classes for which a disjunction exists between their super-classes or between them and their super-classes. These classes are therefore inconsistent. If an inconsistent class exists in the input model, or if there is a cycle in the inheritance relation, an invariant violation is detected and showed by the state view.

```
B> class :
     {(CLASSEXPRESSION1,ENTITY1), (CLASSEXPRESSION2,ENTITY3),
      (CLASSEXPRESSION3,ENTITY4), (CLASSEXPRESSION4,ENTITY1),
      (CLASSEXPRESSION5,ENTITY2), (CLASSEXPRESSION6,ENTITY3),
      (CLASSEXPRESSION7,ENTITY1), (CLASSEXPRESSION8,ENTITY2)}
B> subClassExp :
     {(AXIOM6,CLASSEXPRESSION3), (AXIOM7,CLASSEXPRESSION5),
      (AXIOM1,CLASSEXPRESSION1)}
B> superClassExp:
     {(AXIOM6,CLASSEXPRESSION4), (AXIOM7,CLASSEXPRESSION6),
      (AXIOM1,CLASSEXPRESSION2)}
B> (subClassExp ; class)~ :
     {(ENTITY2,AXIOM7), (ENTITY4,AXIOM6), (ENTITY1,AXIOM1)}
B> (superClassExp ; class) :
     {(AXIOM1,ENTITY3), (AXIOM7,ENTITY3), (AXIOM6,ENTITY1)}
B> inheritance :
     {(ENTITY2,ENTITY3), (ENTITY1,ENTITY3), (ENTITY4,ENTITY1)}
B> Entity_name :
     {(ENTITY1,"Plant"),(ENTITY2,"Animal"), (ENTITY3,"Fauna_Flora"),
      (ENTITY4,"Carnivorous_Plant")}
```

Fig. 8. Meeduse interactive debugger.

4.3   Pattern SubClassOf Without Inconsistency

Once the logical properties of the pattern are defined (and debugged), the specification of the OWL evolution pattern can be done. Our intention is to provide
support beyond simple inconsistency detection, which is already possible in our
approach as illustrated in Figs. 3 and 4. The resulting inconsistency is tagged
with an annotation conforming to the guidelines of [6]. A more pragmatic approach is to define patterns by means of proved operations that do not accept
invariant violations, which is more relevant to the application of a formal approach. Indeed, the advantage of a formal method is to keep the application in
a safe state space during its execution. When the user requires a change that
matches the pattern, the tool directly performs one of the alternatives suggested
by the pattern, which keeps the ontology consistent.

Figure 9 shows the initial ontology (left side) and all possible change requirements (right side) including the definition of Carnivorous_Plant as a sub-class
of Animal. The user can select the alternative he is interested in. In this case
there are two alternatives to apply the SubClassOf pattern between classes Car-
nivorous_Plant and Animal:

- The selected one (marked with "SubClassing") generates a hybrid subclass
  and leads to the ontology of Fig. 1a, and
- The last one (marked with "SuperClassing") generates an hybrid subclass and
  produces the ontology of Fig. 1b.

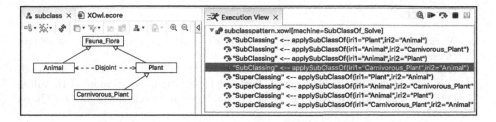

Fig. 9. Pattern SubClassOf without inconsistency

Our formal specification of the SubClassOf pattern without inconsistency has been proved correct with AtelierB, meaning that we are guaranteed that its execution preserves both the structural properties of the OWL language and the considered logical properties. All our artifacts are publicly available at: https://github.com/meeduse/Samples/tree/main/xowl.

5 Conclusion and Perspectives

Managing ontology evolution while maintaining consistency is still a challenge, inspite of approved efforts in ontology design and validation. Several process models are proposed in the literature to capture the key steps of an evolution task. Nonetheless, most of them deal with the methodological aspects of the change. This work contributed to this field via a formal executable DSL approach built on the Meeduse language workbench. The originality is that it combines together MDE and Formal Methods, and applies their strengths to the ontology evolution problem. As far as we know, mixing both paradigms together to ensure a correct pattern-driven change management, has not been investigated before. In fact, the Xtext DSL deals with the structural properties of the ontology, and the B specifications include its logical properties. This formal framework allowed us to prove the correctness of the change patterns.

Our approach is focused on change application and validation step during an ontology evolution process. Since the usage of the B method is transparent for the end-user, we believe that the approach can be extended to manage complex changes (combination of elementary changes) and ontology population (instances). Another perspective, on which we are actively working, is ontology inconsistency resolution. In our opinion, an "intelligent" recommendation mechanism that repairs automatically the ontology or suggests the best resolution actions is required in case of inconsistency.

References

1. Abrial, J.R.: The B-Book: Assigning Programs to Meanings. Cambridge University Press, New York (1996)
2. Aït-Ameur, Y., Ait-Sadoune, I., Hacid, K., Mohand Oussaid, L.: Formal modelling of ontologies within Event-B. In: First International Workshop on Handling IMPlicit and EXplicit Knowledge in Formal System Development, Xi'an, China (2017)
3. Bah, M.O., Boudi, Z., Toub, M., Wakrime, A.A., Aniba, G.: Formalizing ontologies for AI models validation: from OWL to Event-B. In: 15th International Conference on Semantic Computing, pp. 455–462. IEEE (2021)
4. Bettini, L.: Implementing Domain-Specific Languages with Xtext and Xtend, 2nd edn. Packt Publishing (2016)
5. Castano, S., et al.: Ontology dynamics with multimedia information: the boemie evolution methodology. In: Workshop on Ontology Dynamics (2007)
6. Djedidi, R., Aufaure, M.-A.: $ONTO\text{-}EVO^{A}$ L an ontology evolution approach guided by pattern modeling and quality evaluation. In: Link, S., Prade, H. (eds.) FoIKS 2010. LNCS, vol. 5956, pp. 286–305. Springer, Heidelberg (2010). https://doi.org/10.1007/978-3-642-11829-6_19
7. Djedidi, R., Aufaure, M.A.: Ontology evolution: state of the art and future directions. In: Ontology Theory, Management and Design: Advanced Tools and Models (2010). https://doi.org/10.4018/978-1-61520-859-3.ch007
8. Hacid, K., Ait-Ameur, Y.: Strengthening MDE and formal design models by references to domain ontologies. A model annotation based approach. In: Margaria, T., Steffen, B. (eds.) ISoLA 2016. LNCS, vol. 9952, pp. 340–357. Springer, Cham (2016). https://doi.org/10.1007/978-3-319-47166-2_24
9. Hannou, F.-Z., Atigui, F., Lammari, N., Cherfi, S.S.: SafecareOnto: a cyber-physical security ontology for healthcare systems. In: Strauss, C., Kotsis, G., Tjoa, A.M., Khalil, I. (eds.) DEXA 2021. LNCS, vol. 12924, pp. 22–34. Springer, Cham (2021). https://doi.org/10.1007/978-3-030-86475-0_3
10. Horrocks, I., Patel-Schneider, P.F.: Reducing OWL entailment to description logic satisfiability. J. Web Semant. **1**(4), 345–357 (2004)
11. Horrocks, I., Patel-Schneider, P.F., van Harmelen, F.: From SHIQ and RDF to OWL: the making of a web ontology language. Web Semant. Sci. Serv. Agents World Wide Web **1**(1), 7–26 (2003)
12. Idani, A.: Meeduse: a tool to build and run proved DSLs. In: Dongol, B., Troubitsyna, E. (eds.) IFM 2020. LNCS, vol. 12546, pp. 349–367. Springer, Cham (2020). https://doi.org/10.1007/978-3-030-63461-2_19
13. Leuschel, M., Butler, M.: ProB: an automated analysis toolset for the B method. Softw. Tools Technol. Transfer (STTT) **10**(2), 185–203 (2008)
14. Obeo: Xtext/sirius - integration, the main use-cases (white paper). https://www.obeodesigner.com/resource/white-paper/WhitePaper_XtextSirius_EN.pdf
15. Riaño, D., et al.: An ontology-based personalization of health-care knowledge to support clinical decisions for chronically ill patients. J. Biomed. Inform. **45**(3), 429–446 (2012)
16. Sadoun, D., Dubois, C., Ghamri-Doudane, Y., Grau, B.: From natural language requirements to formal specification using an ontology. In: 25th IEEE International Conference on Tools with Artificial Intelligence, pp. 755–760. IEEE CS Press (2013)
17. Sirius: http://www.eclipse.org/sirius/. Accessed 07 Oct 2020
18. Zablith, F., et al.: Ontology evolution: a process-centric survey. Knowl. Eng. Rev. **30**(1), 45–75 (2015)

Generating SPARK from Event-B, Providing Fundamental Safety and Security

Asieh Salehi Fathabadi$^{(\boxtimes)}$, Dana Dghaym, Thai Son Hoang,
Michael Butler, and Colin Snook

ECS, University of Southampton, Southampton, UK
{a.salehi-fathabadi,d.dghaym,t.s.hoang,m.j.butler,cfs}@soton.ac.uk

Abstract. Event-B is a formal method that facilitates rigorous analysis and correct-by-construction development of software and hardware systems. SPARK is a computer programming language for the development of high integrity software. Linking Event-B at design level and SPARK at implementation level allows us to formally verify the relationship between application-level requirements and software implementations. Event-B is supported by an integrated development environment, Rodin, and extension plug-in tools, enabling various validation and verification techniques. However it lacks a comprehensive code generation feature with support for data structures, to connect to implementation. In this paper, we propose a tool to translate verified Event-B models into the SPARK programming language. We describe the translation rules and how the proposed tool can be integrated with other EMF-based plug-ins in Rodin. We demonstrate the proposed translation rules through a 'smart ballot box' case study.

Keywords: SPARK · Formal methods · Event-B · Security · Safety

1 Introduction and Motivation

SPARK[1] is a subset of the Ada programming language targeted at safety and security to create highly reliable software. SPARK restrictions ensure that the behaviour of a SPARK program is simple enough that formal verification tools can be applied to prove the absence of runtime errors, such as arithmetic overflow, buffer overflow and division-by-zero. The SPARK language and toolset for formal verification have been applied over many years to aircraft systems, control systems and rail systems.

However, we believe that safety and security are fundamental design goals. To expand the boundaries of safe and secure application, formal proving techniques can be applied at an earlier design stage. This result brings additional confidence in the software by capturing all the possible bugs before application development. The most recent version DO-178C of the avionics certification standard allows using both tests and proofs as acceptable verification methods [11].

---

[1] https://www.adacore.com/about-spark.

© The Author(s), under exclusive license to Springer Nature Switzerland AG 2022
P. Fournier-Viger et al. (Eds.): MEDI 2022, CCIS 1751, pp. 179–192, 2022.
https://doi.org/10.1007/978-3-031-23119-3_13

Using formal modelling and refinement from system requirements to design level, provides rigorous evidence of safe and secure system specification. The process results in a high-integrity, precise specification. However, to translate this into an implementation without introducing an error-prone manual step, code generation tools are needed.

Event-B [2] is a method to formally model and verify safety- and security-critical systems. The supporting tool set, Rodin [3], provides a comprehensive environment for modelling, validation and verification techniques, such as theorem proving and model checking. However there is no support for generating SPARK from the Event-B verified model including programming data structures like records. In this paper, we propose a set of translation rules from the Event-B formal design model to the SPARK programming level and outline how the translation could be implemented as a plug-in for the Rodin tool set. The translation rules support formal and automatic transformation of Event-B components and elements, including our record data structure extension to Event-B, into the SPARK language.

Our motivation includes providing the system level verification, by applying the formal methods, alongside the verification of software component, provided by the SPARK verification feature. Our case study, the Smart Ballot Box (SBB) for ensuring only legitimate and valid ballots can be cast, demonstrates how attacks at the abstract level can be verified by applying the Event-B method prior to SPARK development. For example, the formal specification for the action of casting a paper, cast_paper event, is guarded by safety and security guards:

event cast_paper refines cast_paper
any paper where
@typeof−paper: paper ∈ papers
@not−already−cast−paper: paper ∉ cast_papers
@not−already−expired: paper_time(paper) ≥ current_time − expiry_duration
@not−already−spoiled: paper ∉ spoiled_papers
@copy−not−already−cast: paper_voter(paper) ∉ paper_voter[cast_papers]
@copy−not−already−spoiled: (∀sp · sp ∈ spoiled_papers ⇒ paper_voter(paper
) ≠ paper_voter(sp) ∨ paper_vote(paper) ≠ paper_vote(sp) ∨
  paper_time(paper) ≠ paper_time(sp))
@not−invalid−paper: paper ∉ invalid_papers
@not−illegitimate−paper: paper ∉ illegitimate_papers

The event guards must hold to enable the occurrences of an event. Here cast_paper event is guarded to ensure that the cast paper is issued from a legitimate source and it is valid, i.e., it has not expired or a copy of it has already been cast or spoiled. The attackers' actions to create or duplicate a paper are specified as events and add the attackers' paper to a set of illegitimate papers. Later, the cast_paper event prevents casting an illegitimate paper. There are a set of Proof Obligations (PO) generated and discharged to ensure the safety and security properties, specified as invariants in the SBB Event-B model. Also INV POs ensure that events, for example the cast_paper event, preserve the safety and security conditions specified as invariants. More details about the formal model can be found in [6]. A consistent SPARK specification can be automatically generated from the verified SBB system model following our proposed translation rules.

The paper is structured as follows. Section 2 discusses related works including previous SPARK generation. Section 3 provides background knowledge of SPARK, Event-B and a short summary of our case study. Section 4 describes the translation rules in detail and outlines the application of some of the rules in our case study. Section 5 outlines a proposed new Rodin plug-in as a supporting tool. Section 6 concludes and outlines future directions.

2 Related Work

The SBB case study has previously been analysed for security using Event-B by Dghaym et al. [6], however, this study did not go beyond the system level requirements. Here we extend the case study by generating SPARK as a first step towards a verified implementation. Conversely, there are examples [9] of research on using SPARK for high integrity, security-critical implementations which do not cover the system level analysis of requirements. Our contribution is to provide the link between these two aspects by deriving formally verified requirements and generating SPARK output to facilitate a high-integrity implementation.

In [15] the authors proposed generating SPARK specifications from Event-B. This initial work did not generate SPARK implementation files and did not support the translation of Event-B record data structures. Here we take a fresh approach to the problem but utilise some ideas from [15]. Similarly, in [12], the authors proposed to generate sequential programs including scheduling events using some merging rules from [2]. The authors also present some patterns for translating simple data structures and formulae. Our work here focused on a richer data structure based on records and generating individual procedure specification and implementation for each Event-B event. It means that the generated procedures can be (later) scheduled differently from what is permitted by the merging rules in [2].

Integration of executable UML models with automatic generation of SPARK has been considered by several authors [5,13,16]. These approaches have similar aims to ours in making a link from modelling to implementations. However, although UML can provide some precision and validation/verification processes via simulation, it does not have the grounding of a mathematically manipulable language and therefore cannot provide formal analysis to the extent of Event-B's theorem provers. Also UML focuses on design and does not support a strong notion of proven refinement from system level requirements.

A formal approach to developing a semantics of the SPARK language in Coq is given by Zhang et al. [17] This is useful to provide a grounding for formally defined tools. Language-specified run-time errors that are relevant for all programs in the language can be contrasted with application-specific run-time errors that correspond to violations of a program's application-specific requirements. Zhang et al [17] address the former notion.

3 Background

3.1 SPARK

SPARK is a specialized subset of Ada [4] designed for engineering high-reliability applications. Ada is a general-purpose language, like C++ or Java, supporting the usual features of modern programming languages. What sets Ada apart from other general-purpose languages is that it was designed from the start with reliability, safety, and security in mind. Ada is used in domains where the correctness of software is critical: space, avionics, air-traffic control, railway, and military.

SPARK facilitates the use of formal methods, so that correctness of software or other program properties can be guaranteed with mathematics-based assurance. Therefore, SPARK is used in the same domains as Ada, by those who value the strong guarantees offered by formal methods. SPARK utilises the strengths of Ada while trying to eliminate all its potential ambiguities and insecure constructs. SPARK programs are by design meant to be unambiguous, and their behaviour is required to be unaffected by the choice of Ada compiler. These goals are achieved partly by omitting some of Ada's more problematic features (such as unrestricted parallel tasking) and partly by introducing additional contracts (to support data dependencies, information flows, state abstraction and data and behaviour refinement) which encode the application designer's intentions and requirements for certain components of a program.

The GNAT toolset[2] also generates a set of verification conditions. These conditions are used to establish whether certain properties hold for a given subprogram. At a minimum, the GNAT prover will generate verification conditions to establish that all run-time errors cannot occur within a subprogram, such as: array index out of range, type range violation, division by zero and numerical overflow.

3.2 Event-B

Event-B [2] is a refinement-based formal method for system development. The mathematical language of Event-B is based on set theory and first order logic. An Event-B model consists of two parts: *contexts* for static data and *machines* for dynamic behaviour. Contexts contain carrier sets s, constants c, and axioms $A(c)$ that constrain the carrier sets and constants. Machines contain variables v, invariant predicates $I(v)$ that constrain the variables, and events. In Event-B, a machine corresponds to a transition system where *variables* represent the states and *events* specify the transitions.

An event comprises a guard denoting its enabling-condition and an action describing how the variables are modified when the event is executed. In general, an event e has the following form, where t are the event parameters, $G(t, v)$ is the guard of the event, and $v := E(t, v)$ is the action of the event: $e == \textbf{any } t \textbf{ where } G(t,v) \textbf{ then } v := E(t,v) \textbf{ end}.$

---

[2] https://www.gnu.org/software/gnat/.

An Event-B model is constructed by making progressive refinements starting from an initial abstract model which may have more general behaviours and gradually introducing more detail that constrains the behaviour towards the desired system. This is done by adding or refining the variables of the previous abstract model and modifying the events so that they use the new variables.

Event-B is supported by the Rodin tool set [3], an extensible open source toolkit which includes facilities for modelling, verifying the consistency of models using theorem proving and model checking techniques, and validating models with simulation-based approaches.

3.3   Smart Ballot Box (SBB)

The main function of the Smart Ballot Box (SBB) [1] is to inspect a ballot paper by detecting a 2D barcode, decode it and evaluate if the decoded contents verifies the paper from a Ballot Marking Device (BMD). If the ballot is valid, then it can be cast into the storage box. Otherwise, the SBB rejects the paper. The SBB does not conduct a full-scale analysis of the document, nor record the choices of the voters, nor tabulate the votes of the ballots it scans. The key function of the SBB is to ensure that only valid countable summary ballot documents that can be tabulated later are included in ballot boxes.

4   Translation Rules and SBB Application

In this section, we present the detailed translation rules, categorised as the type of source element, from the Event-B elements to the SPARK elements. Also we outline application of some of the translation rules to the Event-B model of the SBB. We assume that we have already refined the model from an abstract system level until it represents the functionality of the proposed design within its environment. That is, our refinements have introduced all the required behaviour and all the entities that will interact with the proposed component that represents the SPARK code. However at this stage state the functionality may still be represented using abstractions beyond typical programming data structures. We start by decomposing the model to separate the software part which we want to generate code for, from other parts of the model related to the environment. In our case study we need to generate the SBB code but not the parts of the model that represent entities that interact with the code. We apply shared variable decomposition [14], where we categorise the events as SBB events or environment events. In order to translate the Event-B model into SPARK we need a straightforward correspondence between the data represented in the model and the data constructs used in SPARK. We therefore refine the decomposed SBB model closer to the implementation constructs of SPARK. For example if a set should be implemented as an array, we need to refine the model to address this implementation choice, by introducing indexing and array size. In the case of arrays we data refine the set to a total function where the domain of the function is an integer range and the range is the array type.

4.1   Component Translation

TR_mchn: The last refined machine in Event-B:

machine mch_name sees ctx_name

will generate the specification SPARK file, *ads*:

with ctx_name; use ctx_name;
package mach_name
with SPARK_Mode => On
end mach_name;

and the implementation SPARK file, *adb*:

package body mch_name with SPARK_Mode is ...
end mach_name;

TR_cntx: Each context in Event-B:

context cntx_name extends ext_cntx

will generate a specification SPARK file, *ads*, and will see all contexts it extends and their extended contexts:

with ext_cntx; use ext_cntx;
package cntx_name
with SPARK_Mode => On
end ctx_name;

4.2   Constant Translation

TR_cnst_non_fun: This rule applies to all constants other than functional relations. The constant type and value are specified as axioms:

constants const_name
axioms
const_name ∈ const_type
const_name = const_value

A non-functional constant is translated into a constant declaration in SPARK:

const_name : constant const_type := const_val;

TR_cnst_fun A constant specified as a function in an Event-B context:

constants cnst_name
axioms cnst_name ∈ dom → ran

is translated into a function in SPARK where the domain becomes the function argument and the range is the return type:

function cnst_name (p_dom : in dom) return ran;

In Event-B, functions may be left abstract (i.e. the constant value is not given). In this case, the SPARK function body has to be implemented manually.

If the domain of a functional constant is defined as a cross product, the SPARK function has multiple arguments corresponding to the cross product:

axioms cnst_name ∈ dom1 × dom2 → ran

then

function cnst_name (p_dom1 : in dom1; p_dom2 : in dom2) return ran;

Below is the application of this rule in SBB:

axioms	function MACAlgorithm
@typeof−MACAlgorithm : MACAlgorithm ∈ TIME × CYPHER_TEXT × KEY → MAC	(t : in TIME ; k : in KEY; c : in CYPHER_TEXT) return MAC;

The left column presents the Event-B specification of the MAC algorithm as a function; While the right column presents the translation of it into a SPARK function.

4.3   Variable Translation

TR_var: Event-B variables in a machine are translated into global variables in SPARK. Variables in SPARK should be initialised, hence the translation of variables in Event-B will require the typing invariant and the initialisation action. The variable var_name specified as below:

variables var_name
invariants @inv_type: var_name : var_type
event INITIALISATION then @act_init: var_name := init_val

is translated to:

var_name: var_type := init_val;

Application of this rule in SBB is as follows:

variables cast_count invariants @cast_count_type: cast_count ∈ ℕ event INITIALISATION then @init_cast_count: cast_count := 0	cast_count : Integer := 0;

TR_var_fun: Arrays in Event-B are specified as function variables:

variables var_name
invariants @inv_type: var_name ∈ Integer−range → arr_type
event INITIALISATION then @act_init: var_name := init_constant_fun

and are translated to the arrays in SPARK:

type var_name is array (Integer−range) of arr_type;
var_name : var_name := init_constant_fun;

In the SBB model, array of cast vote are specified as a function variable as below left, and is translated to the SPARK array as below right:

invariants @cast_arr_type: cast_arr ∈ 0..max_votes−1 → BARCODE event INITIALISATION then @cast_arr_init: cast_arr := (0..max_votes−1) × {null_barcode}	type barcode_array is array (0 .. Max_Votes−1) of BARCODE; cast_arr : barcode_array := (others => null_barcode);

4.4   Record Translation

TR_record: A record in Event-B [8]:

record rec_name
field_name : field_type

is translated into a record in SPARK:

type rec_name is
field_name : field_type ;
end record;

Below is the application of this rule in SBB to specify the barcode record:

record BARCODE
   paper_time : TIME
   paper_encrypted_ballot
   : CYPHER_TEXT
   paper_mac: MAC
	type BARCODE is record
	   paper_time : TIME;
	   paper_encrypted_ballot
	   : CYPHER_TEXT;
	   paper_mac : MAC;
	 end record;

4.5   Event Translation

When translating an event in Event-B, we are looking at the last refinement level, where the model is closer to the implementation level. Event translation includes rules for the corresponding parameters, guards and actions of the event, as described in this section.

TR_event: A non-initialisation event in Event-B:

event evt_name

is translated into a procedure in the specification SPARK file *ads*:

procedure evt_name;

and a procedure in the implementation file *adb*:

```
procedure evt_name is
  begin
  end evt_name;
```

Initialisation event translation is performed as part of variable translation (TR_var).

TR_par_input: This rule applies to input parameters of events in an Event-B machine. Type of a parameter is defined as a guard of the corresponding event:

```
event evt_name any p1 p2 where
@grd_type_p1: p1 ∈ p1_type
@grd_type−p2: p2 ∈ p2_type
```

The event input parameters are translated to procedure parameters of the corresponding event in SPARK:

```
procedure evt_name(p1 : in p1_type; p2 : in p2_type)
```

TR_par_output: In Event-B, in most cases, event parameters are input parameters. However there are cases that parameters are used to represent a return value. Event-B events do not return values; parameters are used to represent the output and the result is specified as a guard. In this special case we can use a special naming convention, *result*, to identify output parameters:

```
event evt_name any p result where
@grd_type_p: p : p_type
@grd_type_r: result : r_type
@grd_value_r: result = r_value
```

When translating to SPARK, in the specification SPARK file *ads*:

```
procedure evt_name(p : in p_type; result : out r_type)
```

and in the implementation SPARK file *adb*:

```
procedure evt_name(p : in p_type; result : out r_type) is
  begin
  result = r_value
  end evt_name;
```

TR_grd_noVar: Guards of an event in the Event-B:

```
event evt_name where
@grd1: grd1
@grd2: grd2
```

are translated into procedure preconditions of the corresponding event in SPARK:

```
procedure evt_name with
Pre => ( grd1
  and then grd2 ));
```

TR_grd_var: If a guard of an event was a function of variables:

```
event evt_name where
  @grd1: grd1(v1)
  @grd2: grd2(v2)
then
  @act1: act1(v1)
```

then it is translated to the input and output of the corresponding procedure in SPARK:

```
procedure evt_name with
Global => (Proof_In => (v2),
In_Out => (v1));
```

and in the *adb* file:

```
procedure evt_name is
  begin
  end evt_name;
```

If global variables are only used in the event guards then they are considered as input only, if the global variables are also updated in the actions then they are considered as input and output of the procedure.

TR_action: Actions of an event in the Event-B:

```
event evt_name then
  @act1: act1(v1)
  @act2: act2(v2)
```

are translated into procedure post-conditions of the corresponding event in the specification SPARK file *ads*, below left. Actions are also translated as the procedure body in the implementation file *adb*, below right:

```
                                           procedure evt_name is
       procedure evt_name with            begin
       Post => (act1(v1)                   act1(v1);
       and then act2(v2));                 act1(v2);
                                           end evt_name;
```

Next is the application of the presented rules in this section, to translate the cast_paper event. The event cast_paper presented in Sect. 1 will be data refined as follows; The cast_papers variable representing the set of all cast papers will be replaced by its array representation cast_arr. cast_arr is defined using the gluing invariant cast_arr $\in 0..\text{max\_votes}-1 \nrightarrow$ papers where max_votes is a natural number representing the maximum number of votes.

```
event cast_paper refines cast_paper any paper i size where
  @typeof-paper: paper ∈ barcode
```

@not−already−expired : paper_time(paper) ≥
current_time − expiry_duration
@grd_not_future : paper_time(paper) ≤ current_time
@mac−check : paper_mac(paper) =
MACAlgorithm(paper_time(paper) ↦ paper_encrypted_ballot(paper)
↦ MACKey)
  @copy−not−already−cast : paper_encrypted_ballot(paper) ∉
  paper_encrypted_ballot[ran(cast_arr)]
  @copy−not−already−spoiled : ∀sp · sp ∈ ran(spoiled_arr) ⇒
  paper_encrypted_ballot(paper) ≠ paper_encrypted_ballot(sp)
      ∨ paper_time(paper) ≠ paper_time(sp)
  @index_type : i ∈ 0 .. max_votes−1
  @grd_size : size = card(cast_arr)
@grd_size2 : size < max_votes
@grd_i_size : i = size
theorem @thrm : paper ∉ ran(cast_arr)
  @grd_i : i ∉ dom(cast_arr) // To discharge gluing inv
then
  @update−cast_papers : cast_arr(i) := paper
end

After preparing the model to a level closer to implementation we can run the code generator which will invoke the related transformation rules, as shown below. The guards for cast_array will be transformed to preconditions in the procedure cast_paper, but to simplify the guards we translate each guard into a Boolean expression function in Spark, which can be reused by different procedures.

```
procedure cast(paper : in barcode) with
  Global => (Proof_In => (spoiled_arr, curr_time, spoil_count),
  In_Out => (cast_arr, cast_count)),
  Pre => ( Invariants(spoil_count,cast_arr)
    and then not already_cast(paper)
    and then not already_spoiled(paper)
    and then valid_time (paper)
    and then validate_barcode(paper)
    and then cast_count in 0 .. Max_Votes−1),
  Post => (Invariants(spoil_count,cast_arr)
    and then already_cast(paper)
    −− and then cast_arr(cast_count) = paper
    and then cast_count = cast_count' old + 1);
```

In Event-B, we do range Checks which has to be changed to an equivalent quantifier. For example if the case element is in the range we will refine it to an existential quantifier and it will then be translated directly using the **for some** in range. If the element is not in a range then it will be refined to the universal quantifier and it will be directly refined to the **for all** in a range. Note that for some and for all can only be applied to range values hence the refinement should be mindful of that.

The function already_cast(paper), the negation of copy−not−already−cast, will be represented as follows:

function already_cast (paper: in barcode) return Boolean is
(for some i in 0..Max_Votes−1 => paper.paper_encrypted_ballot =
       cast_arr(i).paper_encrypted_ballot);

The guards are only translated into preconditions in the specification, whereas the actions are translated into postconditions in the specification and also as the procedure body in the implementation:

procedure cast(paper : in barcode) is
begin
cast_arr(cast_count) := paper;
cast_count := cast_count + 1;
end cast;

In addition to the guard and action transformation rules, other rules, such as the parameter (Tr_par_input) and event (TR_event) translation rules, are also applied here.

5   Tool Support

We have developed a prototype Rodin plugin for the translation from Event-B to SPARK. The translation uses our translation framework [10] which is a general framework (implemented as Rodin/Eclipse plug-ins) for defining translations between 2 EMF (Eclipse Modelling Framework) meta-models. We already have an Event-B EMF metamodel which we use in our other Rodin plugins. We defined a light-weight SPARK EMF meta-model as the target abstract syntax. This meta-model only captures the "outer" syntax of SPARK representing the structure of the SPARK package specification and body, with respect to composition of the procedure specification and implementation. It does not capture details of the mathematical language of SPARK, such as predicates for specifying preconditions and post-conditions. Based on our existing Event-B EMF meta-model and this SPARK EMF meta-model, we define a translator by providing a set of Java-based rules via an Eclipse extension point mechanism of our EMF-to-EMF translation framework. The implemented rules correspond to the translation rules presented in Sect. 4. The result of the translation is a SPARK model instance in memory which would by default be serialised in EMF's XMI syntax. In order to obtain a serialisation in the usual SPARK concrete syntax we use XText [7] to override the default serialisation. XText is a powerful framework for developing the concrete syntax of programming languages and domain-specific modelling languages. XText relies on EMF for representing the in-memory (abstract syntax) representation of any parsed files. As a result, the model generated by our Event-B to SPARK translation is serialised into the usual, human-readable, SPARK syntax.

6   Conclusion and Future Works

In this work, we proposed to integrate the formal methods, Event-B with the semi-formal SPARK verification technique, to achieve fundamental safety and security. For this purpose, code generation techniques can facilitate the integration of system level verification and implementation level verification. In this paper, we presented the translation rules and associated tool for automatic transformation of Event-B formal method to SPARK language. Also we highlighted how formal modelling/verification can model attacks and ensure safety/security earlier at design level.

We applied the translation rules and generated SPARK to the software part of the SBB model presented in [6]. We could use the environment model to generate a testing harness as suggested in [15]. However, here we focus on the software specification and leave testing for future work. All the events and contexts related to the SBB specification were used to generate the SPARK code and we managed to verify all the SPARK pre/post conditions. Some functions such as the specifics of the encryption algorithm has to be developed manually due to the abstract nature of Event-B.

In future work, we will continue to apply the code generation tool to further safety and security case studies and extend the translation rules where appropriate. For example, we are currently working on the translation of invariants. Another area that warrants further investigation is refinement and knowing when the model is ready for translation. The refinement of data structures may be relevant to this question and, in particular, we are currently developing more comprehensive record refinement techniques and tools. We will also work on formally verifying the soundness of the Event-B to SPARK translation rules, one possibility would be using the semantics of SPARK in Coq [17] and also formalising the semantics of Event-B.

Acknowledgements. This work is supported by the following projects: HD-Sec project, which was funded by the DSbD Programme delivered by UKRI. And HiClass project, which is part of the ATI Programme, a joint Government and industry investment to maintain and grow the UK's competitive position in civil aerospace design and manufacture.

References

1. Galois and Free & Fair. The BESSPIN Voting System (2019). https://github.com/GaloisInc/BESSPIN-Voting-System-Demonstrator-2019. Accessed 16 Aug 2022
2. Abrial, J.R.: Modeling in Event-B: System and Software Engineering. Cambridge University Press, Cambridge (2010)
3. Abrial, J.R., Butler, M., Hallerstede, S., Hoang, T., Mehta, F., Voisin, L.: Rodin: an open toolset for modelling and reasoning in Event-B. Softw. Tools Technol. Transfer **12**(6), 447–466 (2010)
4. Barnes, J.: Bibliography, 2nd edn., pp. 951–952. Cambridge University Press, Cambridge (2022). https://doi.org/10.1017/9781009181358.045

5. Curtis, D.: SPARK annotations within executable UML. In: Pinho, L.M., González Harbour, M. (eds.) Ada-Europe 2006. LNCS, vol. 4006, pp. 83–93. Springer, Heidelberg (2006). https://doi.org/10.1007/11767077_7
6. Dghaym, D., Hoang, T.S., Butler, M., Hu, R., Aniello, L., Sassone, V.: Verifying system-level security of a smart ballot box. In: Raschke, A., Méry, D. (eds.) ABZ 2021. LNCS, vol. 12709, pp. 34–49. Springer, Cham (2021). https://doi.org/10.1007/978-3-030-77543-8_3
7. Eysholdt, M., Behrens, H.: Xtext: implement your language faster than the quick and dirty way. In: OOPSLA, pp. 307–309. ACM (2010). http://doi.acm.org/10.1145/1869542.1869625
8. Salehi Fathabadi, A., Snook, C., Hoang, T.S., Dghaym, D., Butler, M.: Extensible record structures in Event-B. In: Raschke, A., Méry, D. (eds.) ABZ 2021. LNCS, vol. 12709, pp. 130–136. Springer, Cham (2021). https://doi.org/10.1007/978-3-030-77543-8_12 https://eprints.soton.ac.uk/448194/
9. Georgiou, K., Cluzel, G., Butcher, P., Moy, Y.: Security-hardening software libraries with Ada and SPARK - a TCP stack use case. CoRR abs/2109.10347 (2021). https://arxiv.org/abs/2109.10347
10. Hoang, T.S., Snook, C., Dghaym, D., Fathabadi, A.S., Butler, M.: Building an extensible textual framework for the Rodin platform. In: Proceedings of the 7th Workshop on Formal Integrated Development Environment, F-IDE2022, to be published
11. Moy, Y., Ledinot, E., Delseny, H., Wiels, V., Monate, B.: Testing or formal verification: do-178c alternatives and industrial experience. IEEE Softw. 30(3), 50–57 (2013). https://doi.org/10.1109/MS.2013.43
12. Murali, R., Ireland, A.: E-SPARK: automated generation of provably correct code from formally verified designs. Electron. Commun. Eur. Assoc. Softw. Sci. Technol. 53 (2012)
13. Sautejeau, X.: Modeling SPARK systems with UML. In: SigAda 2005, pp. 11–16. Association for Computing Machinery, New York (2005). https://doi.org/10.1145/1103846.1103848
14. Silva, R., Pascal, C., Hoang, T.S., Butler, M.: Decomposition tool for Event-B. Softw. Pract. Experience 41(2), 199–208 (2011). https://eprints.soton.ac.uk/271714/
15. Sritharan, S., Hoang, T.S.: Towards generating SPARK from Event-B models. In: Dongol, B., Troubitsyna, E. (eds.) IFM 2020. LNCS, vol. 12546, pp. 103–120. Springer, Cham (2020). https://doi.org/10.1007/978-3-030-63461-2_6
16. Wilkie, I.: Executable UML and SPARK Ada: the best of both worlds (2005). https://abstractsolutions.co.uk/wp-content/uploads/2018/03/Executable-UML-and-SPARK-Ada-V2.1.pdf
17. Zhang, Z., Robby, Hatcliff, J., Moy, Y., Courtieu, P.: Focused certification of an industrial compilation and static verification toolchain. In: Cimatti, A., Sirjani, M. (eds.) SEFM 2017. LNCS, vol. 10469, pp. 17–34. Springer, Cham (2017). https://doi.org/10.1007/978-3-319-66197-1_2

An Entropy-Based Approach: Handling Uncertainty in IoT Configurable Composition Reference Model (CCRM)

Soura Boulaares[1]([✉]), Salma Sassi[2], and Sami Faiz[3]

[1] National School for Computer Science, Manouba, Tunisia
boulaaressoura@gmail.com
[2] Faculty of law, Economic, and Management Sciences, Jendouba, Tunisia
[3] Higher Institute of Multimedia Arts, Manouba, Tunisia

Abstract. IoT has expanded the boundaries of the world with physical entities and virtual components as a result of the proliferation of published, invoked, and consumed IoT items. Adapting a reference model-based approach that respects the design by reuse or configuration philosophy has become a significant challenge. Hence, according to the life cycle of a connected Thing, in order to make the composition or consumption of an IoT object reusable and configurable a configurable reference composition model (CCRM) is proposed by Atlas+. Considering the complexity, scalability, heterogeneity and dynamic changes of the IoT environment, a composition model reuse will reduce costs, burdens and time spent. However, at the design time, the configurable conception mechanism applied to the composition reference model brings uncertainty related to the choice of the most relevant composition plan. This uncertainty is due to the fact that the configurable model means a restriction of the behaviour represented by an existing composition plan model. This behaviour restriction will allow only one desired composition of the reference model while eliminating unwanted ones. The uncertainty associated in selecting the optimum configuration plan from among the options is the challenge of IoT composition. In this paper, we will propose an entropy-based uncertainty measure that allows us to take into account the dynamic aspect of the model at design time and quantify this uncertainty in order to assess the predictability and efficiency of the composition plan of IoT Objects.

Keywords: Uncertainty · Configuration · Reference model · Entropy measure · Social things · IoT

1 Introduction

The Internet of Things (IoT) is a network of physical devices and software components that communicate and deliver IoT services. IoT applications aim to increase network elements to provide more powerful and value-added services.

© The Author(s), under exclusive license to Springer Nature Switzerland AG 2022
P. Fournier-Viger et al. (Eds.): MEDI 2022, CCIS 1751, pp. 193–206, 2022.
https://doi.org/10.1007/978-3-031-23119-3_14

Major challenges related to the discovery and composition of IoT services need to be overcome for the Internet of Things (IoT) to progress. The data produced by IoT systems is generally large and diverse. As a result, the composition of these services, as well as the appropriate use of these data, are becoming more complex. Therefore, a common or reference model is needed to define all objects and their interactions, as well as to transmit them to high-level knowledge.

In the Internet of Things paradigm, the challenge of service composition is crucial. A composite Thing Service can clearly define how the real-world elements specified as object services called Thing services (TS) [8] interact. Among the major challenges of IoT composition is variability and uncertainty. Variability consists in finding a solution that makes it possible to generalize the composition plane into a common or reference plane. This plan will model all redundant (common) and heterogeneous (variable) parts.

Specifically, in [8] generated a Social Internet of Things(SIoT) configurable Composition Reference Model (CCRM)that is compatible with the IoT context. Since the Social Internet of Things is an extended paradigm of the Internet of Things, it represents the fusion and enrichment of the standards of social networks, social web and IoT. It also allows to model and estimates the relationships between connected objects. In our previous work, we represented a reference model and a language thanks to the semantic rules for configurable composition management. This model handles configurable primitives: Thing Services (TS or cTS), Thing Relationships (TR or cTR) and Recipes (R or cR) (which can be configurable and not configurable). The proposed language also presents configurable operations and connectors (cAND, cOR, cXOR, sequence).

On the other hand, the challenge of uncertainty has been treated according to several visions. Some deal with the uncertainty of the data used during the composition such as in the IoT context [2,3] and in the web context [1,5]. In addition, others have treated the suitability of a composition plan as a safe plan [1] problem or as a problem of navigating between web resources [5].

In [9,11] they prepared an uncertainty measure based on entropy in the business process context (BPs). In 1865, Rudolf Clausius introduced the concept of entropy in thermodynamics as a measure of the transmission of heat between two solid entities at different temperatures. Its objective is to assess the degree of disorganization or unpredictable nature of a random variable. Attempt to persuade the system to deviate from its normal behaviour. Quantification of variability and uncertainty is an important benefit because systems with high uncertainty cannot promise better predictions.

However, the uncertainty related to the composition reference model was not addressed. This uncertainty is related to the best choice of composition design among all possible designs in the reference model. This challenge has been addressed in the context of traditional business process (BP) to address uncertainty related to the execution of process activities. In addition, uncertainty was addressed in the context of configurable business processes to address uncertainty when designing a configurable business process reference model (cBP). By analogy, to the business process, a composition plan is conceived as a workflow or a process of composition of services.

Therefore, in order to estimate the uncertainty related to the configurable composition reference model, we will rely on the approaches applied on the BP and cBP [9,11].

- **Motivation example:**
  The configuration of a composition plan model (CCPM)in the context of IoT [8] means the restriction of the compositions represented by the latter so that it allows only the desirable composition of the model. In contrast, all undesirable compositions are blocked or deleted. Since a configurable composition reference model (CCRM) consists of configurable primitives, operators and connectors (TS, TR, R, Operations and Connectors) [8] which have several configuration possibilities, the user will choose a path from several according to his needs. Because of the number and unpredictability of options, choosing from all possible configurations might be difficult. Since a composition reference model includes several types of configurable points (data services their relations and operations and connectors), we are interested in the uncertain choices related to configurable connectors used in a composition reference model. Our main objective is to be able to generalize uncertainty management and make it applicable to any configurable composition language of IoT objects. Generally, the choice is taken arbitrarily generally or depends on user preferences. As a result, there is a need for transparency in uncertainty management. Taking the example of a configurable composition process made by our Atlas+ language [8] in Fig. 1. This section of the Reference Model will represent several alternatives that do not have weight to help choose the best option. Assuming that a configurable control relationship will take as a precondition a set of configurable Thing Services connected by the configurable cXOR connector. The choice of individual connectors is arbitrary. Generally to choose a possibility one either applies the XOR between the TS, or the sequence and one of the possibilities is taken as indicated in Fig. 1.

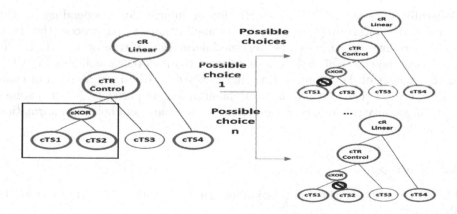

Fig. 1. The possible choices of a CCPM.

– *The question will be to know which option among those present possibilities can be chosen?*
On one hand, the reference model of the configurable composition is of immense importance since it allows to integrate of several variants of the composition, and this allows to give a general vision to the companies that the individuals who use the model. But, on the other hand, the increasing number (scalability) of connected objects and the diversity of their use, the variability of context and dynamic changes have created a major challenge:
– *Uncertainty in the Configurable Composition Reference Model:*
This challenge consists in the variability of the choice of the most pertinent variant in a composition plan (different planes and possible composition path). As a result, there is a lack of pertinence when it comes to selecting the appropriate variable composite variant.

The main objective of this paper is to propose an approach to measure the uncertainty of the Configurable CRM. Our approach allows us to handles:

– variable composition processing using the Atlas+ language.
– the decomposition of global CCRM into fragments and generating metadata on the CCRM to model the imperfection of the CCRM.
– the quantification of CCRM uncertainty using the generated data and the Shannon entropy applied to the points of variation, especially the connectors.

The remaining of this paper is organised as follows; Sect. 2 presents the basic background, Sect. 3 depicts the state of the art works, Sect. 4 handles an overview of our approach and the final section consists of implementation and discussion.

2   Background

2.1   Uncertainty

Generally, uncertainty is an imperfection of information. Depending on the domain, the uncertainty depends on its application, if you process the data, the uncertainty determines whether the information is true or false [5,10]. It consists of the choice of path is more relevant than another possible path [9,11]. In the context of data composition, it consists of the best aggregation of composite services [1,5]. According to a classification proposed by [4], it assumes that uncertainty generalizes incompleteness, vagueness and belief in information or knowledge.

2.2   Entropy

The uncertainty of the information is usually calculated by the entropy of Shannon:

$$H(X) = \sum_{i=1}^{n} P(x_i) \, u(u_i) = -K \sum_{i=1}^{n} P(x_i) \, log_2 P(x_i)$$

Where X is a random variable taking possible states $(x_1, x_2, ..., x_n)$ with probabilities($P(x_1)$, $P(x_2)$, $\cdots$,$P(x_1)$)The entropy H(X) is the expectation of u(xi) which is xi's uncertainty. In the context of configurable business processes, the uncertainty measurement is modelled by four formulas in relations to the configurable connector. Used to estimate the uncertainty of cBP at the time of design [9,11]. As follows the list of The uncertainty measurement formulas of cBP [11]:

- **Sequence:** $U(seq) = \sum_{i=1}^{I} (U(a_i))$
- **Andc:** $U(And) = \sum_{i=1}^{n} (U(B_i))$
- **Orc:** $U(Orc) = -\sum_{i=1}^{N} \frac{1}{3 \times ((2^n - 1) - 2n)} \, log_2 \frac{1}{(2^{n-1})} - \sum_{i=1}^{N} \frac{1}{3 \times ((2^n - 1) - 2n)} \times U(B_i)$
- **Xorc:** $U(Xorc) = -\sum_{i=1}^{N} \frac{1}{(2^{n-1})} \, log_2 \frac{1}{(2^n - 1)} - \sum_{i=1}^{N} \frac{1}{(2^n - 1)} \times U(B_i)$

2.3 Atlas+ Configurable Composition Language

In this study, we'll use the Atlas+ [8] programming language. Atlas+ is a language supporting the configurable composition for IoT and in particular for the SIoT. Atlas+ is made up of primitives known as recipes, services, and relationships. Operations like evaluate are intended to assess primitives. Additionally, services are filtered using the filter operator according to preferences based on Thing relationships. Another Atlas+ language component that considers the connectors used to accumulate services at the level of a relationship and services and at the level of the recipe. It should be noted that all Atlas+ components can be either classic or configurable.

3 Related Works

A review of the literature shows that the uncertainty of configurable composition has not been managed over the years, as a solution for managing uncertain reference model for IoT configurable composition. On the other hand, uncertainty has been modelled in the context of traditional composition and at the level of business processes. It is assumed that the semantics of a business process is close to the concept of composition, since composition consists of a workflow of several services that will be executed to achieve a common goal. Similar to a business process, the composition goes through a design before the model is executed. In fact, some work has dealt with uncertainty in traditional and configurable business processes. In the work of [9] the authors estimated the uncertainty related to the performance of conditional tasks within a traditional business process. In [11] the authors estimated the uncertainty associated with designing configurable business processes. Where multiple process alternatives may be considered relevant. In this context, the uncertainty measure is used to determine which of the process path choices is the most relevant. In terms of the composition of services in the context of the Internet of Things only in [2,3] the authors have addressed the uncertainty of composition based on the approach of possible worlds. In [5] we have addressed the navigation and the uncertain composition of web resources were addressed in the context of web-related data. Another work was proposed in [7] which is based on the same language in [5,6] to model the uncertainty of Health WoT Data. In [1] authors have proposed an approach based on possible worlds (database uncertainty model). They proposed formulas and algorithms for achieving the uncertain composition of web services. In addition, they checked the safety of the classic composition model with a "safe plan" algorithm. [8] have proposed a configurable model and reference language for the composition of IoT services. This model is based on semantic rules that describe the primitives and operations for the evaluation of these primitives. In addition, the language handles configurable connectors. In the context of cBP and IoT services [12] the authors have proposed an approach for configurable allocation management of IoT services at a configurable business process level.

4   Uncertain Entropy Measure Approach of the IoT CCRM

4.1   Overview of the General Architecture

In Fig. 2 we present the general architecture of the computation of the uncertainty measurement of the configurable composition reference model (CCRM). This architecture is composed of four modules, each of which we detail in the following sections. The first module consists of configurable modelling of SIoT Objects we use as Atlas+ language. Then the result of the first module will be

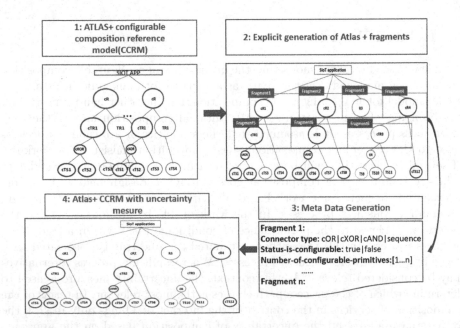

Fig. 2. General architecture of the uncertainty measure computation process

the entry of the next module. The second module consists of an explicit fragmentation of reference model for this we use the algorithm of fragmentation of XML file into a set of fragments. Then we generate the metadata that contributes to the uncertainty calculation such as: **Type of configurable connectors;** **Status:** configurable or not; **Number of configurable nodes or primitives:** relative to each Atlas+ fragment. Therefore we will have an annotated Atlas+ CCRM. Finally, we measure the uncertainty of the generated sub-fragments from the lowest level of the tree (the TS/cTS) until the arrival at the root (cR: upper level). In the next sections, we will detail each of these steps.

4.2   Atlas+ IoT CCRM

A configurable composition plan reference model (CCRM) is an oriented graph with two essential components: nodes and links [8]. The nodes can be, for the atlas+ language, the primitives; Thing Service (cTS/TS), Thing Relationship (cTR/TR), Recipe (cR/R) and operations or connectors such as OR, Or exclusive (XOR) and AND. On the other hand, a configurable composition plan model is a reference model consisting of several variants of the same composition plan. The example in Fig. 3 illustrates a configurable composition plan reference model using Atlas+ (Configurable Atlas) as a modeling language. Our language [8] assumes that a SIoT application is a classic or configurable recipe sequence (R and cR). And conversely, a recipe is a fragment of the IoT application. A composition reference model consists of one or more recipes. And likewise the recipes are composed of several relations which constitute themselves fragments for a recipe. Configurable nodes are modelled with bold lines. Primitives can be present (ON) or absent (OFF) when customizing the model. Our example consists of five primitives (cTS1, cTS2, TS3, TR, R) whose only configurable ones are (TS1, TS2, TR and R). The example also consists of a configurable connector(XOR). The following link depicts the CCRM:[1]

4.3   Fragments Generation and Modelling

For this proposed step, we will use the XML parsing technique that groups the CCRM under the Atlas+ language. This parser takes a CCRM input and calculates its tree. In fact, this technique is based on a unique, modular, linear time decomposition of a linear composition process graph into a hierarchy of sub-graphs (fragments) with a single Input and a single Output. The first step, consist of the decomposition into sub-fragments using the parser that scans the Atlas+ primitives. Although Atlas+ represents the implicit generation of fragments (primitives). Figure 4 shows a fragmented CCRM with its Atlas+ tree thanks to the proposed XML parsing algorithm. The second step consists of the organization of the fragments: during this stage the fragments resulting from the Atlas+ fragmentation can be organized in the form of a tree structure. Each fragment becomes a node.

---

[1] https://github.com/SouraBoulaares/CCRM.git

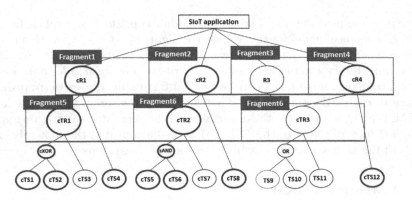

Fig. 3. Step1: Configurable composition reference model Atlas+

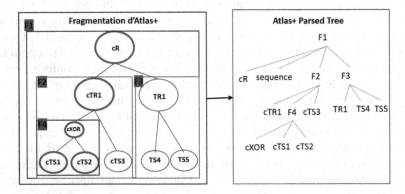

Fig. 4. Step2: Fragmented Atlas+ with the parser

4.4 Meta-Data Generation

In this step, the Meta-Data parser is concerned about generating metadata on fragments created in the previous step. We will use as input the output of the CCRM fragmentation which is the atlas+ tree structure. Afterwards, we generate the specific Meta-Data about each fragment (F1, F2, F3 and F4). Thus, we will have as a result a configurable Atlas+ tree. This tree is the essential data for the CCRM uncertainty computation. Figure 5 illustrates the possible metadata for the fragment tree after completion of the first and second steps.

4.5 Uncertainty Modelling and Computation

The uncertainty measurement of each fragment is based on its specific Meta-Data. In this section, we propose four uncertainty computation formulas for the four possible cases (sequence, cAND, cXOR, cOR).

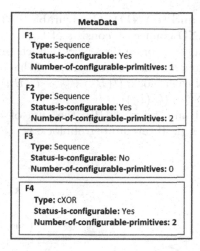

Fig. 5. Step 3: metadata generation

Case 1: cSequence. This is the set of primitives executed in sequence. A sequence may contain configurable and non-configurable primitives. Figure 6 shows a sequence of classical primitives (TR/TS/R).

U (classical primitive) $= 0$; (the value 0 means that there is no uncertainty) Or U: Uncertainty. As shown in Fig. 6, the configurable primitive can be present or absent in the sequence depending on the configuration, so that the probability that it is present or absent is determined by the $\frac{1}{2}$ ($\frac{1}{2}$ present or $\frac{1}{2}$ absent) value. As we have already detailed in our previous paper [8] any configurable primitive can be enabled or disabled for a sequence-type recipe or a sequential recipe.

U (configurable primitive) $= \frac{-1}{2}log_2(\frac{1}{2})$

Hence, **U (cSequence)** $= \sum_{i=1}^{I}(U(cprim_i))$

With (cprimi: configurable primitive i, U(prim i): uncertainty of the primitive i, I: total number of primitives).

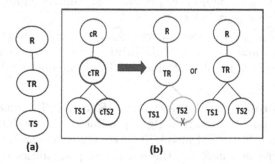

Fig. 6. classic sequence (a) and configurable sequence (b)

Case2: Configurable and (cAND). Configurable AND (cAND) is configured in classic AND. The AND connector consists of two or more parallel branches (in the case of a relationship or service block). Each branch forms a block of primitives (Bi), so the uncertainty measurement of this connector is the sum of the uncertainty value of each block; $U(B1) + \cdots + U(Bn))$ as depicted in Fig. 7.

So, $\mathbf{U(cAND)} = \sum_{i=1}^{n} (U(B_i))$

With (Bi: Block Bi and 1 i n, U(Bi): the sum of the uncertainty of the primitive elements of Bi, n: Total number of blocks).

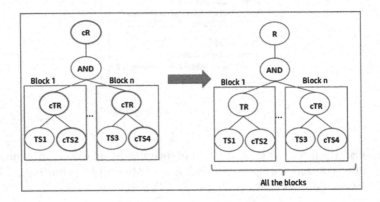

Fig. 7. The configurable AND connector

Case3: Configurable XOR (cXOR). The configurable XOR (cXOR) can be configured in sequence or in conventional XOR. This connector consists of two or more branches and among these multiple branches. Where, one and only one will be executed. We will therefore have $2^n - 1$ (-1 because we must have at least one branch) and therefore the transition probability will be $\frac{1}{2^n-1}$ as in Fig. 8. The configuration possibilities of two primitives in the case of a cXOR are as follows:

cXOR(Prim1, Prim2) = XOR(Prim1, Prim2, XOR (Prim1, Prim2)). We have n=2; we then have 3 configuration possibilities. And therefore, the uncertainty of a cXOR connector is as follows:

$\mathbf{U(cXOR)} = -\sum_{i=1}^{N} \frac{1}{(2^n-1)} log_2 \frac{1}{(2^n-1)} - \sum_{i=1}^{N} \frac{1}{(2^n-1)} \times U(B_i)$

with (Bi:Block Bi and $1 \leq i \leq n$, U(Bi):the sum of the uncertainties of the elements of Bi, n:Total number of blocks,N:Total number of sub-blocks).

Case4: Configurable or Connector (cOR). A cOR connector consists of sequences with n possibilities. And three possible connectors(OR, AND, XOR). Each connector has $2^n - 1 - 1$ possibilities and since the case of the sequence has already been taken into account, the number of possibilities of each connector becomes. The total number of possibilities is:

$((n)) + 3 \times (2^n - 1 - n)) = 3 \times (2^n - 1) - 2n$ as presented in Fig. 9.

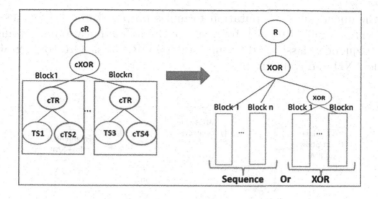

Fig. 8. The configurable XOR connector

The probability of transition will be $\frac{1}{3 \times (2^n - 1) - 2n}$

Hence, $\mathbf{U(cOR)} = -\sum_{i=1}^{N} \frac{1}{3 \times ((2^n - 1) - 2n)} log_2 \frac{1}{(2^{n-1})} - \sum_{i=1}^{N} \frac{1}{3 \times ((2^n - 1) - 2n)} \times U(B_i)$

with (Bi: Block Bi and $1 \leq i \leq n$, U(Bi): the sum of the uncertainties of the elements of Bi, n: Total number of blocks, N: Total number of sub-blocks).

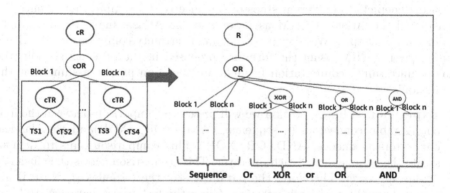

Fig. 9. Configurable OR connector

5   Implementation

This part consists in providing the global prototype architecture Uncertain Measure of configurable composition Reference Model (UMCCRM) for computing uncertainty in the CCRM. This tool is based on two main steps:

- the metadata of each fragment of the IoT application to compute the uncertainty from the bottom up to the root of the tree that contains the entire CCRM.

– and the uncertainty computation formulas proposed in the previous section
that will be applied to each fragment of the IoT application according to its
type (sequence, classic XOR, configurable XOR, classic OR, configurable OR,
classic AND and configurable).

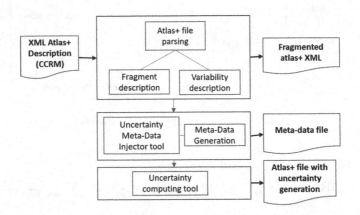

Fig. 10. UMCCRM

Figure 10 depicts the different steps of the uncertainty computation. Hence, we
considered the Atlas+ CCRM as input and the Atlas+ file with uncertainty
generation as output. We have proposed an uncertainty computation Algorithm
1 of a given CCRM using the formulas proposed in the previous section (the
classic uncertainty computation formula are based on [9] (we assume that the
execution's U = 0).

– **Experiments:** In this section we will present a comparison between the four
configurable connectors (c sequence, cAND, cOR, cXOR)and the classical
connectors (sequence, AND, OR, XOR). This comparison aims to demonstrate the complexity of each operator. This comparison shows the efficiency
of our approach. Therefore, the parameters of the formulas (I, N, n, i) are
changed equally with each iteration. Given the lack of comparative study on
configurable connectors. We decided to compare our results with those deduce
[9]. The findings are set out in Fig. 11.
– **Discussion:** The uncertainty values interpreted by entropy are translated as
follows: If U = 0 means that the IoT application fragment of any connector is
certain; If the value of uncertainty tends towards infinity then the uncertainty
increases; If the uncertainty value tends to 0 then the uncertainty is lower. In
the sequence and classic AND framework the values are always set to O. For
classic connectors OR is the most uncertain and complex and always marks
the highest values. For the case of configurable connectors by comparing them
to all other connectors, we note that cOR is the most complex and brand all
the time the peak follows it the cXOR and so on.

Algorithm 1. Uncertainty CCRM computation

Require: *state, Type, N, Atlas + fragmentedfile*
Ensure: *U//uncertaintymeasure*
 1: U=O   ▷ For each fragment in the CCRM
 2: **switch** *Type* **do**   ▷ U of classic connectors in the CCRM
 3:   **case** *sequence*
 4:     **Assert** (*seq_Fromula*())
 5:   **case** *AND*
 6:     **Assert** (*and_Fromula*())
 7:   **case** *OR*
 8:     **Assert** (*or_Fromula*())
 9:   **case** *XOR*
10:     **Assert** (*xor_Fromula*())
11:   **case** *csequence*   ▷ U of configurable connectors in the CCRM
12:     **Assert** (*cseq_Fromula*())
13:   **case** *cAND*
14:     **Assert** (*cand_Fromula*())
15:   **case** *cOR*
16:     **Assert** (*cor_Fromula*())
17:   **case** *cXOR*
18:     **Assert** (*cxor_Fromula*())
19: **EndCase**
20: **return** *U*

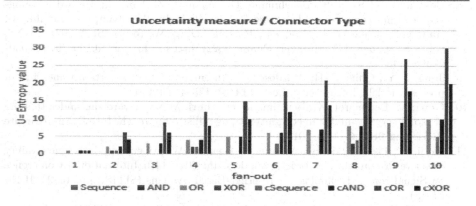

Fig. 11. Uncertainty comparison

6   Conclusion and Future Work

We propose an entropy-based metric to assess uncertainty in IoT composition in general and in particular configurable SIoT models to measure the effectiveness of predictability. The proposed measure allows the estimation of configurable composition uncertainty related to the design time. This helps configurable composition designers to improve composition processes so they are more predictable, less complicated, less error-prone, and easier to understand.

References

1. Amdouni, S., Barhamgi, M., Benslimane, D., Faiz, R.: Handling uncertainty in data services composition. In: 2014 IEEE International Conference on Services Computing, pp. 653–660. IEEE (2014)
2. Awad, S., Malki, A., Malki, M.: Composing wot services with uncertain and correlated data. Computing **103**(7), 1501–1517 (2021)
3. Awad, S., Malki, A., Malki, M., Barhamgi, M., Benslimane, D.: Composing wot services with uncertain data. Futur. Gener. Comput. Syst. **101**, 940–950 (2019)
4. Bouchon-Meunier, B., Nguyen, H.T.: Les incertitudes dans les systèmes intelligents. Presses universitaires de France (1996)
5. Boulaares, S., Omri, A., Sassi, S., Benslimane, D.: A probabilistic approach: a model for the uncertain representation and navigation of uncertain web resources. In: 2018 14th International Conference on Signal-Image Technology & Internet-Based Systems (SITIS), pp. 24–31. IEEE (2018)
6. Boulaares, S., Sassi, S., BenSlimane, D., Faiz, S.: A probabilistic approach: uncertain navigation of the uncertain web. Concurr. Comput. Pract. Exp. **34**, e7194 (2022)
7. Boulaares, S., Sassi, S., Benslimane, D., Faiz, S.: Uncertain integration and composition approach of data from heterogeneous WoT health services. In: Gervasi, O., Murgante, B., Hendrix, E.M.T., Taniar, D., Apduhan, B.O. (eds.) ICCSA 2022. LNCS, vol. 13376, pp. 177–187. Springer, Cham (2022). https://doi.org/10.1007/978-3-031-10450-3_13
8. Boulaares, S., Sassi, S., Benslimane, D., Maamar, Z., Faiz, S.: Toward a configurable thing composition language for the SIoT. In: Abraham, A., Gandhi, N., Hanne, T., Hong, T.-P., Nogueira Rios, T., Ding, W. (eds.) ISDA 2021. LNNS, vol. 418, pp. 488–497. Springer, Cham (2022). https://doi.org/10.1007/978-3-030-96308-8_45
9. Jung, J.-Y., Chin, C.-H., Cardoso, J.: An entropy-based uncertainty measure of process models. Inf. Process. Lett. **111**(3), 135–141 (2011)
10. Omri, A., Benouaret, K., Benslimane, D., Omri, M.N.: Towards an understanding of cloud services under uncertainty: a possibilistic approach. Int. J. Approximate Reasoning **98**, 146–162 (2018)
11. Saidi, M., Tissaoui, A., Benslimane, D., Benallal, W.: An entropy-based uncertainty measure of configurable process models. In: 2018 14th International Conference on Signal-Image Technology & Internet-Based Systems (SITIS), pp. 16–23. IEEE (2018)
12. Suri, K., Gaaloul, W., Cuccuru, A.: Configurable IoT-aware allocation in business processes. In: Ferreira, J.E., Spanoudakis, G., Ma, Y., Zhang, L.-J. (eds.) SCC 2018. LNCS, vol. 10969, pp. 119–136. Springer, Cham (2018). https://doi.org/10.1007/978-3-319-94376-3_8

A Maude-Based Rewriting Approach to Model and Control System-of-Systems' Resources Allocation

Charaf Eddine Dridi[1,2](✉) , Nabil Hameurlain[1] , and Faiza Belala[2]

[1] LIUPPA Laboratory, University of Pau, Pau, France
{charaf-eddine.dridi,nabil.hameurlain}@univ-pau.fr
[2] LIRE Laboratory, Constantine 2 University – Abdelhamid Mehri,
Ali Mendjeli, Algeria
{charafeddine.dridi,faiza.belala}@univ-constantine2.dz

Abstract. Systems-of-Systems (SoSs) are increasingly used to integrate and execute numerous Constituent-Systems (CSs) to offer new functionalities that cannot be offered by its CSs. However, designing well-tuned SoS to deal with a variety of control issues such as resources management and temporal constraints violations while providing high-level assurance about their specified behavior is very challenging. Thus, due to the lack of explicit models for their resources allocations and quantitative features, SoSs executions are becoming more complex and cannot be effectively controlled. To solve these problems, we propose a generic metamodel to control resources and behavioral features. The proposed approach deals with different resources of SoSs and provides control actions in order to manage their structural, temporal and behavioral aspects. The latter are grouped in a single model holding SoSs executions according to specific quantitative needs. One step further, the specifications of the proposed metamodel are encoded using Maude language that makes it possible to analyze various requirements needed by CSs.

Keywords: SoS · Missions · Resources allocation · Dynamic behavior · Maude

1 Introduction

In the last decade, Systems-of-Systems (SoSs) appeared as new software technologies that integrate a set of various subsystems from different subfields, and which offer a reliable and more natural alternative to build an emerging vision for the next generation of large-scale systems [1,2]. These systems are designed to integrate multiple independent and functional Constituent-Systems (CSs) into a larger system in several important application domains. Consequently, an SoS has the ability to offer new functionalities to users that cannot be offered by its CSs, but emerge from their combination.

© The Author(s), under exclusive license to Springer Nature Switzerland AG 2022
P. Fournier-Viger et al. (Eds.): MEDI 2022, CCIS 1751, pp. 207–221, 2022.
https://doi.org/10.1007/978-3-031-23119-3_15

When referring to the period needed to execute a mission, duration is considered as a simple temporal constraint between the mission starting (resp. ending) instants. Since we consider the duration of missions as a generalization of almost all other constraints [3–5], the challenge here is that the time constraints applied on missions can be implicitly affected by the physical environmental features of different constituents. These features are related to resource properties. The resources that SoS requires at run-time, are sometimes limited, unlimited, renewable or not renewable [14–16]. As a solution, we use a set of features that specify different resources for local and global CSs. These features are the basis of developing a dependencies-based approach that has sets of attributes describing the status of the resources that the CSs may use concerning availability, consumption, (non)shareability, disruption, renewability, and withdrawal. Furthermore, these challenges are not easy to manage, i.e. independent evolution and dynamic resources' allocation may cause these CSs to behave differently, for instance, they may affect the missions executions and the CSs interactions and communications within the SoS.

In this paper, we provide a formal modeling approach that reduces the complexity of designing SoS temporal, resources and interaction behavior. We adopt an approach to provide a metamodel for specifying resources properties, and temporal aspects of SoSs. We focus on defining the logical structure and behavior of qualitative and quantitative features involved in the SoS definition, which allows describing all the SoS features at the same high level of abstraction. Hence, to create this generic model, we have to consider a set of concepts, aspects, and features, i.e. hierarchical composition of CSs, missions organization, roles interactions, temporal variations, resources allocation, etc. Employing such a metamodel is a promising mean to show its ability to initiate our model into a wide range of specific technologies, and to provide more holistic solutions in SoSs applications. To instantiate our metamodel, our choice was oriented towards the Maude language, thanks to its expressivity, it can execute, and validate all the relevant specifications of our proposed metamodel without losing information and features. Its executable semantics support and boolean expressions are relevant to design the states predicates of each CS. The language offers more accurate modeling of SoSs whose behavior depends on Missions quantitative time/resources. It uses the equations of rewriting theory to specify the data types of all components. It also offers a model-checker engine that provides a symbolic state-based verification of SoS properties. The proposed Maude-based SoS modeling approach is one of the effective solutions that enable the conditional rewriting rules to describe the unexpected behavior of SoSs. In this specification, we present a way to combine different Maude Modules at different entities levels to produce global knowledge about the entire SoS design.

2   Related Work

We represent in this section some relevant studies that analyze different SoS aspects. Adopting a SysML visual modeling language, the authors of [6,7]have proposed a semi-formal SoS conceptual model which serves as a domain-independent vocabulary for SoSs.

The authors of [8,9] have introduced a formal approach based on Architectural Description Language, that inspires its syntax notation from Bigraphs to model hierarchical structures of CSs and their roles. They have focused on modeling the potential cooperation between CSs by offering a syntax-based description that manages a set of constrained events and roles links affecting the global mission of an SoS.

The authors of [10] have proposed B3MS for SoSs modeling based on the formal technique of Bigraphical reactive systems with an inspiring vision from multi-scale modeling. They have given a method to address the dynamic aspect of SoS by providing model-based rules of basic reconfigurations. They have relied on bigraphical reaction rules to only express the different scales of configurations at the levels of composition, communications nodes, etc.

In [11], the authors have investigated the interplay between SoS and CSs architectures. The approach includes both the design of an SoS architecture that considers the architecture of an existing CS, and the architecting of CSs so that they can later become constituents of an SoS. However, they lacked concrete features that can govern CSs operating in an autonomic manner in a constrained environment. The work presented in [12] have focused on developing a metamodel that represents SoS ontologies. It can be used for both modeling activities and ontology definitions. Even so, to support complete and systematic analysis and design of an SoS.

The authors of [13] have presented a hybrid method based on service-oriented and ontology-based requirements for SoSs modeling. Initially, they introduced an SOA conceptual model, then, modeling semantics through defining multiple ontologies which is important for domain knowledge reuse is defined.

We take advantage of some interesting coordination models presented in [14–16] to build up our generic model for SoSs. The latter addresses the issue of conflicts over different types of resources and categorization during their consumption by missions or production by SoS roles. However, our proposed model is governed by a set of behavioral constraints controlling the execution. SoSs may evolve using transition systems, by using specific actions which are conditioned by specific predicates stating whether the application of the action is allowed or not. The metamodel semantic is implemented using Maude language, offering enough expressiveness and simulation with its executable rewriting logic.

3   A Metamodeling Approach for SoSs Architecture Description

In this section, we provide a generic model that introduces different features.

3.1   Principle: A Generic Metamodel for SoSs Architecture

An SoS consists of a set of CSs that are playing different roles in an organizational and technical environment. These roles jointly realize a common goal.

More precisely, an SoS model is composed of several missions or goals that execute together, to offer one global mission. These missions may need resources for execution such as humans, machines, services, etc. Moreover, SoSs can take four different types [2], Directed (with a central managed purpose and central ownership of all CSs), Virtual (lack of central purpose and central alignment), Collaborative (with voluntary interactions of CSs to achieve a goal), and Acknowledged (independent ownership of the CSs). Subsequently, the proposed definition is introduced based on the key consideration that SoSs are more than simply a set of connected CSs sharing data and offering missions, but, it defines the logical structure and behavior of qualitative/quantitative features involved in Directed SoSs. Therefore, the aim is to define a generic metamodel that will be used as a basis for the modeling and designing of these specific types, whose CSs can have their operational/managerial independence but their emergent behavior is subordinated to a specific mission and Central Controller of all CSs. In this context, we are interested in modeling and specifying public resources

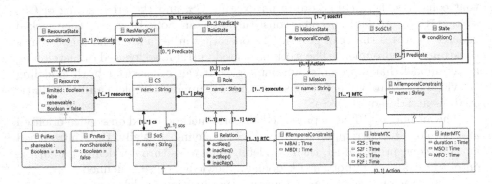

Fig. 1. General components of SoS metamodel.

(PubRes) which are shareable and consumed by missions within the SoS, and private resources (PrvRes) which are not shareable and consumed by missions within the CS. They could also be characterized by some other properties, i.e. limited, unlimited, renewable, and not renewable, as well as whether they are logical or physical resources. Therefore, Resources Management Controller consists of specifying the behavior to be adopted to manage the consumption and the production techniques in the SoS. It consists of a set of actions that are triggered in case the specified triggering conditions are fulfilled. The consumption relationships between missions and resources (resp. production relationships between roles and resources) in our specification are based on specific features that they have. We consider it as an SoS-specific refinement of model property-based resource consumption cycles introduced in [14,15], the proposed controller supports the consumption/production by the matching of specific Missions/Roles states with the features of the different resources. Other important characteristics that we should take into account are: Mission duration, Roles interactions, Desired/ Unwanted behavior, Composition, and CSs configuration [17].

The reason of proposing this metamodel is the need to cope with the variety of the features, and to offer a comprehensive specification to design any SoS architecture. Thus, the proposed metamodel is specified by enhancing the concepts with the essential quantitative features that an SoS should have. Therefore, we need to provide a generic metamodel for SoS design, to be able to show the ability of the metamodeling approach in instantiating architectural models of such systems and to explicitly represent all features. The specification level (see the five classes on top, red rectangle in Fig. 1) concerns concepts related to the definition of behavior controllers (ResMangCtrl and SoSCtrl) that can manage the different states of Resources, Missions and CSs. Their Precondition Predicate associations represent the states conditions that must hold before and be fulfilled after a transition Action of a Resource, a Mission, or a CS.

3.2 Motivating Example

An example of EmergencyIncidentManagementSoS [18], abbreviated EIMSoS, will be considered to concretize the basic contributions made by our approach. As an illustrative example of a critical SoS, consider a collection of autonomous and interacted CSs tasked with transporting and evacuating trauma patients and persons injured in emergencies. To accomplish this Mission, the considered EIMSoS must be designed in such a way that the CSs can perform some missions that cannot be provided by any CS. The visual representation of the EIMSoS elements is shown in Fig. 2. A general idea of this SoS is that, after receiving an emergency call, the CallCenterSoS(CCSoS) which provides the ComputerAided-Dispatch(CAD) as a public resource containing all the necessary information and the data of the incident, directly interacts with its two sub-call centers CSs i.e. AmbulanceStation(ASCS) and the RetrievalServices (RSCS). The three systems have access to shared resource CAD and then, they start switching between different roles (e.g. Receiver and/or Dispatcher) to accomplish a set of local missions, i.e. they discuss the available decisions and determine the optimal response after the emergency's call.

If the decision is made to dispatch a helicopter and/or an ambulance, the Dispatcher will activate the role Aero-Transport and/or GroundEmergency of AeroMedical-EmergencyCS (AMECS) and GroundEmergencyCS (GECS) to start performing the corresponding missions, i.e. preparing the medical teams, flight, departure, first aid, and evacuating, etc. At this stage the ambulance and the helicopter are considered as private (PrvRes) and limited resources provided by AMECS and GECS, respectively. Upon arrival at the hospital, the Hospital-SoS(HSoS) will interject and hands-over to the corresponding wards (e.g. LabCS, RadioCS, etc.) that can accomplish the necessary missions (e.g. triage assessment, diagnosis, lab tests, etc.) and take care of the patient.

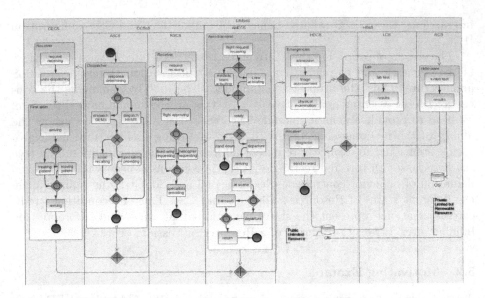

Fig. 2. BPMN model for emergency incident management SoS (EIMSoS).

4   An Executable Maude-Based Specification of SoSs Metamodel

We have chosen the class diagram notation to represent graphically the proposed metamodel. UML Classes, aggregations, inheritance, etc. are used to represent elements and entities of this metamodel. In this section, we give an overview of our proposed object-oriented modules, and how they enable the implementation of the different entities and concepts mentioned in the previous metamodel; and how to model and control their emergent behavior. On the one hand, implementing these entities enables the designers to easily understand the SoSs architectures, including reasoning about their features by providing a high-level model of their structures. On the other hand, it enables them to instantiate it by specifying several mappings or transformations to deduce various executable models. To this end, we use Maude as a logical basis that can provide a clear definition of the object-oriented semantics and makes it a good choice for the formal specification in the form of a transition system.

The following subsections introduce the encoding of our metamodel in Maude. The latter is chosen as a formal specification language and verification platform, because it is expressive enough for specifying all the concepts, aspects and features of the metamodel. On the one hand, it gives a clear specification of the object-oriented notations classes in the metamodel (e.g. SoS, CS, Mission, Roles, Ressource classes, etc.), and describes all the relevant specifications without losing information and features. And on the other hand, it also offers several rewriting-based theories that are adequate for simulating the interdependencies (i.e. the Predicate and Action associations) between these classes by controlling the states of each Mission, Resource and CS, and offering more accurate modeling of SoSs whose behavior depends on Missions properties.

4.1 Motivating the Use of Maude

The declarative language Maude [19] is a very expressive equational and rewriting logic language, as well as one of the most powerful languages in programming, executable formal specification, and formal analysis and verification. Its computation consists of a logical deduction by concurrent rewriting modulo the equational structural axioms of each theory. Maude is an object-oriented modeling language for the specification of distributed systems. Our specification is structured using two types of combined Maude modules, i.e. Real-Time and Object-Oriented ones. Each module consists of one specific class with a set of attributes, and it can import other specifications using that can provide the basic data types like objects, and configurations. What interests us is that Maude employs asynchronous message passing and, thus, the rewrite rules specify the possible transitions that may happen in different entities in SoSs. We specify the behavioral transitions using the following rewrite rules [19, 20]:

$$crl[R] : \{t\} => \{t'\} \ in \ time \ u \ if \ (condition)$$

The triggering conditions of these rules R are state predicates of different entities in the metamodel and their triggered actions are encoded as Maude functional computation. These rules describe how to trigger an action, and how the corresponding control states are transited from one to another. It rewrites the left-hand term "t" of the rule into its right-hand term t' on which an action or a transition is applied. A conditional rewrite rule is triggered when its specified (condition) is satisfied. The idea is that in concurrent CSs forming an SoS, the concurrent states, which are called a configuration, consist of a multiset of objects and messages. Rewrite rules then define transitions between such configurations.

4.2 SoS Global Structures and Behaviors Modeling

In this section, we present Maude specification modules and classes that encode the different concepts mentioned in the metamodel including the necessary information to describe their Predicates, Action, States, and the overall behavior emerging from their relationships, structures and features.

Modeling Missions and Their Durations

Structure Encoding. The Mission specification structure, the checking predicates, the temporal constraints and their violations signals are specified in the system module MissionSpec. The latter contains all the necessary declarations defining the class Mission, which has seven attributes, namely, the starting sm and the ending em instants of a Mission instance m, the allowed completion duration of the mission expressed as an interval dn, dx, the mission state during st, whether its temporal constraints are respected or not sg, and the required resources rs to finish the execution.

$$\textbf{class Mission} \mid : \textbf{Nat, em : Nat, dn : Nat,}$$
$$\textbf{dx : Nat, st : Stt, sg : Sig, rs : Rs}$$

States and Temporal Constraints Predicates Encoding. We use (sort Sig) to model three signal events that are triggered to represent and check the duration constraints i.e. isTCR(M) is triggered to inform that there are no duration violations, isMnV(M) and isMxV(M) to indicate that minimum duration is violated (resp. maximum duration). Additionally, Stt attribute in class Mission is employed to specify different states of the Mission instance (see Fig. 2) For convenience and simplicity, note that in this Fig. 2 and in the upcoming statecharts, all transitions have a set of Predicates that are denoted by letters before the "/", and Actions that are denoted by numbers after it. Thus, we encode in Maude a set of predicates PMis() that represent the conditions (left table) which must be true to trigger the transitions. These predicates will be employed by the Resources Controller to enable (or disable) the different actions AMis()(right table) in a given mission instance. One simple example of how this transition system works: consider a mission M waiting in WaitConsResp state for an authorization from the controller to consume R, (i.e. it has already sent to the controller a Consumption Request using the action (1) AMis). If the controller accepts the request, the predicate (c) PMis = isConsReqAccepted (M, R) becomes true and it will trigger the action (3) AMis = execute(M) to move the M from WaitConsResp state to Executing to state.

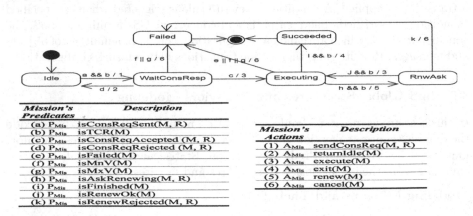

Mission's Predicates	Description
(a) P_{Mis}	isConsReqSent(M, R)
(b) P_{Mis}	isTCR(M)
(c) P_{Mis}	isConsReqAccepted (M, R)
(d) P_{Mis}	isConsReqRejected (M, R)
(e) P_{Mis}	isFailed(M)
(f) P_{Mis}	isMnV(M)
(g) P_{Mis}	isMxV(M)
(h) P_{Mis}	isAskRenewing(M, R)
(i) P_{Mis}	isFinished(M)
(j) P_{Mis}	isRenewOk(M)
(k) P_{Mis}	isRenewRejected(M, R)

Mission's Actions	Description
(1) A_{Mis}	sendConsReq(M, R)
(2) A_{Mis}	returnIdle(M)
(3) A_{Mis}	execute(M)
(4) A_{Mis}	exit(M)
(5) A_{Mis}	renew(M)
(6) A_{Mis}	cancel(M)

Fig. 3. Mission transition system.

Afterward, we use a set of Maude conditioned rules in combination with different monitoring and TC predicates to control and manage the current mission states, as well as its consumption requirements. These rules take the form:

```
crl [name] : action (AMis) < Mid : Mission | state : S > => < Mid :↩
     Mission | state : S' > if (PMis)
```

Modeling Resources Types

Structure Encoding. Using Maude specification, the resource specification in terms of structure, transitions predicates, and features are specified in the module ResourceSpec, it contains all the necessary declarations defining the class Resource. The latter has four attributes, namely, prp, resSt, resT and resC to keep track of, respectively, the resource property prp and its actual state rSt, whether it is a public or private resource via rT attribute and their category.

class Resource | prp : Prop, resSt : ResSt, resT : ResT, resC : ResC.

States Predicates Encoding. We use (sorts Prop ResSt ResT) to express their properties (ul: unlimited, lim : limited and lr : limited but renewable), their states (see Fig. 4), if they were PubRes or PrvRes, and their categories. We identify a set of predicates PRes() and actions ARes() to specify the states and transitions of each resource in Maude. E.g. consider a resource R, available and holding in NotConsumed, the controller will fire the transition "b && c / 2" and perform the action (2) ARes=consume(R) to move to Consumed state whenever the two predicates (b)PRes and (c)PRes become true. These later indicate that there is a consumption approval and R is public. In case R is private the transition system will fire "d / 3", i.e. if (d)PRes=isPrvRes(R) execute (3) ARes=lock(R) and move to state Locked.

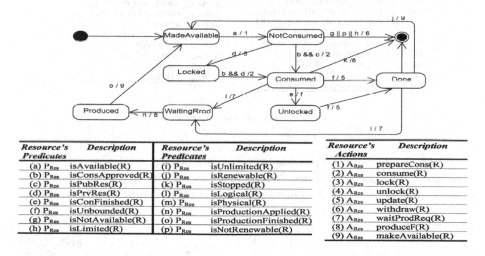

Resource's Predicates	Description	Resource's Predicates	Description	Resource's Actions	Description
(a) P_{Res}	isAvailable(R)	(i) P_{Res}	isUnlimited(R)	(1) A_{Res}	prepareCons(R)
(b) P_{Res}	isConsApproved(R)	(j) P_{Res}	isRenewable(R)	(2) A_{Res}	consume(R)
(c) P_{Res}	isPubRes(R)	(k) P_{Res}	isStopped(R)	(3) A_{Res}	lock(R)
(d) P_{Res}	isPrvRes(R)	(l) P_{Res}	isLogical(R)	(4) A_{Res}	unlock(R)
(e) P_{Res}	isConFinished(R)	(m) P_{Res}	isPhysical(R)	(5) A_{Res}	update(R)
(f) P_{Res}	isUnbounded(R)	(n) P_{Res}	isProductionApplied(R)	(6) A_{Res}	withdraw(R)
(g) P_{Res}	isNotAvailable(R)	(o) P_{Res}	isProductionFinished(R)	(7) A_{Res}	waitProdReq(R)
(h) P_{Res}	isLimited(R)	(p) P_{Res}	isNotRenewable(R)	(8) A_{Res}	produceF(R)
				(9) A_{Res}	makeAvailable(R)

Fig. 4. Resource transition system.

The Maude rules will be employed along with the Controller rules to manage the resource logical and physical behavior. These rules take the following form:

```
crl [name]  : action (ARes) < Rid : Resource| resStt : S > => < Rid :↵
        Resource| resStt : S' > if (PRes) .
```

Modeling Roles Structures and Behaviors

Structure Encoding. The Role specification structure, states and transitions are specified in the system module RoleSpec. The latter contains all the necessary declarations defining the class Role.

class Role | rlSt : RlSt, rclck : Time, mrt : TimeInf, timer : TimeInf.

This class highlights the current state rlSt, local clock (rclck), the time value needed to receive the response message from the role whose (mrt) value is interested, and a timer (timer); four kinds of actions involving roles actReq(R, R'), actRep(R, R'), etc. On the one hand, to specify the resources production behavior provided by different Roles, we encode in Maude a set of predicates PRol() and actions ARol() in a given Role.

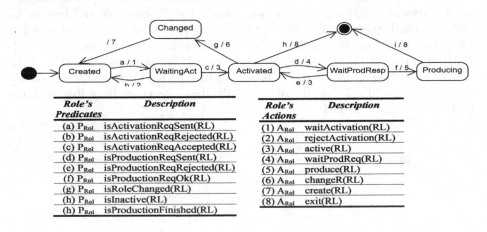

Role's Predicates	Description		Role's Actions	Description
(a) P$_{Rol}$	isActivationReqSent(RL)		(1) A$_{Rol}$	waitActivation(RL)
(b) P$_{Rol}$	isActivationReqRejected(RL)		(2) A$_{Rol}$	rejectActivation(RL)
(c) P$_{Rol}$	isActivationReqAccepted(RL)		(3) A$_{Rol}$	active(RL)
(d) P$_{Rol}$	isProductionReqSent(RL)		(4) A$_{Rol}$	waitProdReq(RL)
(e) P$_{Rol}$	isProductionReqRejected(RL)		(5) A$_{Rol}$	produce(RL)
(f) P$_{Rol}$	isProductionReqOk(RL)		(6) A$_{Rol}$	changeR(RL)
(g) P$_{Rol}$	isRoleChanged(RL)		(7) A$_{Rol}$	create(RL)
(h) P$_{Rol}$	isInactive(RL)		(8) A$_{Rol}$	exit(RL)
(h) P$_{Rol}$	isProductionFinished(RL)			

Fig. 5. Role transition system.

On the other hand, we provide in the same module, an RT specification with a protocol for specifying roles Active/Inactive actions times. To manage the unwanted behavior related to missions-based conflicts using rewrite rules :

```
rl [!Synch—missions]: notSynch(M1, M2) < M1 : Mission | sm : T1, em :↵
        T'1, dn : Dn1, dx : Dx1, st : WaitResp, sg : Sig, rs : PrvR, rC↵
        : RC > < M2 : Mission | sm : T2, em : T'2, dn : Dn2, dx : Dx2, ↵
        st : WaitResp, sg : Sig, rs : PrvR, rC : RC > => < M1 : Mission↵
        | sm : T1, em : T'1, dn : Dn1, dx : Dx1, st : changeSt(WaitResp↵
        , Dx1 — Dn1, Dx2 — Dn2), rs : PrvR, rC : RC > < M2 : Mission | ↵
        sm : T2, em : T'2, dn : Dn2, dx : Dx2, st : changeSt(WaitResp, ↵
        Dx1 — Dn1, Dx2 — Dn2), rs : PrvR, rC : RC >
```

4.3   SoS Controller Behavior Modeling

In this section, we present our controller which supports the Resources Management, dynamic roles interactions, and Systems behavior states.

Modeling Resources Management Controller

The consumption via the combination of Missions and Resources states and the production is supported by matching the states of Roles and Resources. The goal is to decide whether to accept or not Missions consumption (resp. Role production) requests. More specifically, this is accomplished by managing different predicated actions to decide to either change or not the different states.

Structure Encoding. To simulate the controller decisions, we present the module ManagementCtrlSpec for allocation/production control. We describe these specifications in terms of structure, consumption/production predicates, and actions. Missions may have access to resources, and Roles may affect their availability during the production process. the latter are presented in class:

class ResManagementCtrl | mission : Oid, role : Oid,

resource : Oid resMangS : ResMangStt.

Resources Management Behavior. In the same module, we use the previous predicates (resp. actions), those related to Missions PMis(), Roles PRol() and Resources PRes(), (resp. AMis(), ARol() and ARes()) to formalize the new ones PRC() (resp. ARC()) that manage the consumption/production of Resources. To illustrate how this controller RC works, consider that RC is holding in AnalyseConReq, the transition system will fire the transition "b/2" and perform the action (2) ARC to move to ConsReqAccepted state whenever the predicate (b)PRC becomes true. The latter is composed of five predicates (see table in Fig. 6): (a)PMis=a consumption request of M is sent, (b)PMis = its temporal constraints are respected, (a)PRes = resource R is available, (b)PRes = consumption is approved, and (c)PRes = R is public see table in Fig. 3 and Fig. 4, respectively. Whenever these predicates become true, the action (2) ARC in turn will trigger two actions(2) ARes = Consume R and (3) AMis = execute M.

We encode allocation actions as conditional rewrite rules, i.e. their triggering conditions are the state predicates encoded above and their triggered actions are encoded as Maude computation. The different actions express atomic controlling behaviors for resource consumption/production, missions and roles. The conditional rewrite rules in this case will be as follows:

```
crl [rewrite-rule-name] : action (ARC) < Rid : Resource| resStt : RS
    > <: Mission | state : MS, res : Rid > < RCid :
ResManagementCtrl |mission : Mid, resource : Rid, state : RCS >
    => < Rid : Resource| resStt : RS' > <: Mission | state : MS,
    res : Rid' > < RCid : ResManagementCtrl |mission : Mid, resource
    : Rid, state : RCS' > If (PRC)  .
```

So as not to go into too much detail, we give a simplified example of the rule[lock-resource]: If Rid is PrvRes, or renewable, the controller sends a lock(Rid) message to Rid and moves to the Lock state.

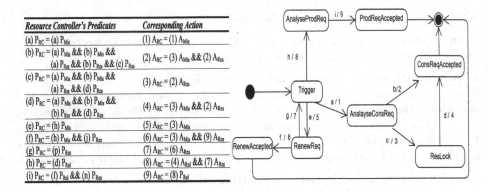

Resource Controller's Predicates	Corresponding Action
(a) $P_{RC} = (a) P_{Mis}$	(1) $A_{RC} = (1) A_{Mis}$
(b) $P_{RC} = (a) P_{Mis}$ && (b) P_{Mis} && (a) P_{Res} && (b) P_{Res} && (c) P_{Res}	(2) $A_{RC} = (3) A_{Mis}$ && (2) A_{Res}
(c) $P_{RC} = (a) P_{Mis}$ && (b) P_{Mis} && (a) P_{Res} && (d) P_{Res}	(3) $A_{RC} = (2) A_{Res}$
(d) $P_{RC} = (a) P_{Mis}$ && (b) P_{Mis} && (b) P_{Res} && (d) P_{Res}	(4) $A_{RC} = (3) A_{Mis}$ && (2) A_{Res}
(e) $P_{RC} = (h) P_{Mis}$	(5) $A_{RC} = (3) A_{Mis}$
(f) $P_{RC} = (h) P_{Mis}$ && (j) P_{Res}	(6) $A_{RC} = (3) A_{Mis}$ && (9) A_{Res}
(g) $P_{RC} = (p) P_{Res}$	(7) $A_{RC} = (6) A_{Res}$
(h) $P_{RC} = (d) P_{Rol}$	(8) $A_{RC} = (4) A_{Rol}$ && (7) A_{Res}
(i) $P_{RC} = (f) P_{Rol}$ && (n) P_{Res}	(9) $A_{RC} = (8) P_{Rol}$

Fig. 6. Resources management controller transition system.

```
crl [lock−resource]: lock (Rid) < Rid: Resource | resSt : PrvRes > < ↵
    Mid : Mission | st : WaitResp, rs : Rid > < ACid : ↵
    AllocationCtrl | Mid, Rid, resSt : AnalyseConsReq > => < Rid: ↵
    Resource | resSt : Locked, rT : PrvRes > < Mid : Mission | st : ↵
    Executing, rs : Rid > < ACid : AllocationCtrl | Mid, Rid, resSt↵
    : ResLock > if (isLock(Rid) && isPubRes(Rid) && Sig =TCR) || (↵
    isLimited(Rid) && isRenwable(Rid) && isAvailable(Rid) && Sig =↵
    TCR) .
```

Systems Behavior Specification

To control the desired behaviors and prevent the unwanted ones, it is necessary to study the global state of the system that could be affected by various factors which are related to the different states of roles, missions, and resource consumption/production. For this aim, the module SoSSpec provides the rewriting logic mean that can specify and check the global configurations of SoS design, and then, ensure the correctness of the evolved SoS substructures.

class SoS | csL : CSL, ml : ML, rl : RL, resl : ResL, csst : CSSt.

Structure Encoding. The SoS Maude-based specification in terms of organizational composition (e.g. ownership, dependency, etc.), transitions, and states are defined in class SoS. The latter has five attributes, namely, mL, rL, resL and CSL, for CSs, missions, roles, and resources lists, and CSSt to describe the CS state (ignored, joined, active, etc.). This class describes the global state of the system according to its CSs.

Systems Global Behavior. We take advantage of the specified states of missions, resources, and roles to describe the behavior of the participating CSs. We give rewrite rules that express guarded configuration implementing the desired behavior of CSs playing Roles.

5   Simulation and Analysis: Case Study 'EIMSoS'

Formal designing of SoSs features depend on different states, which are expressed in terms of the relationship between Missions and Resources, etc.

5.1   Maude-Based Modeling of the Resource Management

We show the use of the Resource Controller by applying a sequence of rules (module ResourceCtrlSpec) on a configuration of the EIMSoS. Given object identifiers, H1, Dep, and AMECSctrl, the following term may represent a configuration with a private and limited resource which is a Helicopter, a mission Departure, a resource controller for AeroMedicalEmergencyCS, respectively, and an access(Dep, H1) message from Dep to AMECSctrl to enable the allocation.

```
< Dep : Mission | 5000, 450, 550, Idle, TCR > < H1 : Resource | lim,↩
     NotCns, PrvRes > < AMECSctrl : AllocationCtrl | Dep , H1 >
```

The execution of this controller is modeled with a set of rewriting rules. For a given scenario, when the mission Dep needs to consume/use the resource H1 the request-analysis rule, will send the controller an access request and it will wait at WaitResp state for a response. If ever H1 is not available (state notAv), the controller will send a rejection (sendRej(H1, Dep)) replying to the mission, and the latter will back to state Idle in request-rejection rule.

```
crl [request−analysis] : < Dep : Mission | 500, 450, 550, Idle, TCR ↩
    > <H1: Resource | lim, resSt, prvRes > accesReq(Dep, H1) => < ↩
    Dep : Mission | 5000, 450, 550, WaitResp, TCR > < H1: Resource |↩
    lim, resSt, prvRes > < AMECSctrl : AllocationCtrl | Dep, H1> ↩
    sendRej(H1, Dep) if resSt (H1) = notAv .
crl [request−rejection] : isReqRej(Dep) < Dep : Mission | 500, 450, ↩
    550, WaitResp, TCR > =>  < Dep : Mission | 500, 450, 550, Idle↩
    , TCR >
```

Contrarily, if the mission receives an ok message (isRespOK (H1, Dep)), it will start the execution (State Exec), and the controller sends a ResLock(H1) message to H1 and moves to the ResLock state.

```
crl [request−accept] : < Dep : Mission | 500, 450, 550, Idle, TCR > ↩
    <H1: Resource | lim, resSt, prvRes > accesReq(Dep, H1) => < ↩
    Dep : Mission | 5000, 450, 550, Exec, TCR > < H1: Resource |lim↩
    , lckd, prvRes > < AMECSctrl : AllocationCtrl | Dep, H1> ResLock↩
    (M, R) if canBeCons(H1) && isTCR(Dep)) || (isLimButRnwable(H1) ↩
    && isAv(H1) && isTCR(Dep)) (lock(H1) && isPubRes(H1) && ↩
    canBeConsumed(H1) && isTCR(Dep)) || (isLimButRnwable(H1) && isAv↩
    (H1) && isTCR(Dep)) .
```

If the mission is taking longer than expected, a signal event MxV(Dep) is transferred to inform the SoS that the duration constraint is violated.

```
crl [request−TCViolated] : < Dep : Mission | ET, 450, 550, Exec, TCR↩
    > <H1: Resource | lim, resSt, prvRes> accesReq(Dep, H1) => <↩
    Dep : Mission | ET, 450, 550, Exec, MxV > < H1: Resource | lim,↩
    lckd, prvRes > < AMECSctrl : AllocationCtrl | Dep, H1> if ET > ↩
    550 .
```

5.2   Maude-Based Modeling of the SoS States Configuration

According to the SoS class, each CSs state has a relation with the list of roles RL that he plays and the relevant list of missions ML as well. Based on these two lists, several states and transitions naturally emerge within a CS that is relevant to the SoS: IgnoredCS, PreparedCS, PassiveCS and ActiveCS.

```
rl [to—ignore] : < RSCS : SoS | ML, RL , csst :CSSt > ignoreCS (RSCS,↩
    RL, ML) => < RSCS : SoS | ML, RL ,IgnoredCS : CSSt > ...
```

6   Conclusion

The main contributions of this paper are threefold, we first proposed a generic metamodel that emphasizes different aspects oriented towards a set of qualitative and quantitative features. Then, we proposed a formal approach for the modeling and specification of SoSs structures and behaviors based on temporal and resources features. Finally, we provided an execution and verification solution of the defined behaviors, using the rewriting-logic-based Maude system, and we validated it using an Emergency Incident Management SoS as a case study.

References

1. Nielsen, C.B., Larsen, P.G., Fitzgerald, J., Woodcock, J., Peleska, J.: Systems of systems engineering: basic concepts, model-based techniques, and research directions. ACM Comput. Surv. **48**(2), 18:1-18:41 (2015)
2. Maier, M.W.: Architecting principles for systems-of-systems. Syst. Eng. **1**(4), 267–284 (1998)
3. Combi, C., Oliboni, B., Zerbato, F.: Modeling and handling duration constraints in BPMN 2.0. In: Proceedings of the Symposium on Applied Computing (2017)
4. Graja, I., Kallel, S., Guermouche, N., Kacem, A.H.: Towards the verification of cyber-physical processes based on time and physical properties. Int. J. Bus. Syst. Res. **13**(1), 47–76 (2019)
5. Cheikhrouhou, S., Kallel, S., Guermouche, N., Jmaiel, M.: Toward a time-centric modeling of business processes in BPMN 2.0. In: Proceedings of International Conference on Information Integration and Web-based Applications & Services (2013)
6. Mori, M., Ceccarelli, A., Lollini, P., Bondavalli, A., Fromel, B.: A holistic viewpoint-based SysML profile to design systems-of-systems. In: 2016 IEEE 17th International Symposium on High Assurance Systems Engineering (HASE) (2016)
7. Mori, M., Ceccarelli, A., Lollini, P., Frömel, B., Brancati, F., Bondavalli, A.: Systems-of-systems modeling using a comprehensive viewpoint-based SysML profile. J. Softw.: Evol. Process. **30**(3), e1878 (2017)
8. Seghiri, A., Belala, F., Hameurlain, N.: A formal language for modelling and verifying systems-of-systems software architectures. Int. J. Syst. Serv.-Oriented Eng. (IJSSOE) **12**(1), 1–17 (2022)
9. Seghiri, A., Belala, F., Hameurlain, N.: Modeling the dynamic reconfiguration in smart crisis response systems. In: 17th International Conference on Evaluation of Novel Approaches to Software Engineering, pp. 162–173. SCITEPRESS-Science and Technology Publications (2022)

10. Gassara, A., Rodriguez, I.B., Jmaiel, M., Drira, K.: A bigraphical multi-scale modeling methodology for SoS. Comput. Electr. Eng. **58**, 113–125 (2017)
11. Axelsson, J., Fröberg, J., Eriksson, P.: Architecting systems-of-systems and their constituents: a case study applying industry 4.0 in the construction domain. Syst. Eng. **22**(6), 455–470 (2019)
12. Baek, Y., Song, J., Shin, Y., Park, S., Bae, D.: A meta-model for representing system-of-systems ontologies. In: Proceedings of the 6th International Workshop on Software Engineering for Systems-of-Systems - SESoS '18 (2018)
13. Zhang, Y., Liu, X., Wang, Z., Chen, L.: A service-oriented method for system-of-systems requirements analysis and architecture design. JSW **7**(2), 358–365 (2012)
14. Halima, R.B., Klai, K., Sellami, M., Maamar, Z.: Formal modeling and verification of property-based resource consumption cycles. In: 2021 IEEE International Conference on Services Computing (SCC), pp. 370–375. IEEE (2021)
15. Maamar, Z., Faci, N., Sakr, S., Boukhebouze, M., Barnawi, A.: Network-based social coordination of business processes. Inf. Syst. **58**, 56–74 (2016)
16. Graiet, M., Mammar, A., Boubaker, S., Gaaloul, W.: Towards correct cloud resource allocation in business processes. IEEE Trans. Serv. Comput. **10**(1), 23–36 (2016)
17. Axelsson, J.: A refined terminology on system-of-systems substructure and constituent system states. In: 2019 14th Annual Conference System of Systems Engineering (SoSE), pp. 31–36. IEEE (2019)
18. Andrews, R., et al.: Leveraging data quality to better prepare for process mining: an approach illustrated through analysing road trauma pre-hospital retrieval and transport processes in Queensland. Int. J. Environ. Res. Public Health **16**(7), 1138 (2019)
19. Meseguer, J.: Rewriting logic and maude: a wide-spectrum semantic framework for object-based distributed systems. In: Smith, S.F., Talcott, C.L. (eds.) FMOODS 2000. IAICT, vol. 49, pp. 89–117. Springer, Boston, MA (2000). https://doi.org/10.1007/978-0-387-35520-7_5
20. Clavel, M., Durán, F., Eker, S., Lincoln, P., Martí-Oliet, N., Meseguer, J., Talcott, C.: The maude 2.0 system. In: Nieuwenhuis, R. (ed.) RTA 2003. LNCS, vol. 2706, pp. 76–87. Springer, Heidelberg (2003). https://doi.org/10.1007/3-540-44881-0_7
21. Author, F.: Article title. Journal **2**(5), 99–110 (2016)
22. Author, F., Author, S.: Title of a proceedings paper. In: Editor, F., Editor, S. (eds.) CONFERENCE 2016, LNCS, vol. 9999, pp. 1–13. Springer, Heidelberg (2016). https://doi.org/10.10007/1234567890
23. Author, F., Author, S., Author, T.: Book title, 2nd edn. Publisher, Location (1999)
24. Author, A.-B.: Contribution title. In: 9th International Proceedings on Proceedings, pp. 1–2. Publisher, Location (2010)
25. LNCS Homepage. http://www.springer.com/lncs. Accessed 4 Oct 2017

Author Index

Printed in the United States
by Baker & Taylor Publisher Services